FINE STRUCTURE
OF CELLS
AND TISSUES

By KEITH R. PORTER, PH.D.

and MARY A. BONNEVILLE, PH.D.

Department of Molecular, Cellular and Developmental Biology
University of Colorado,
Boulder, Colorado

with the collaboration of
SUSAN A. BADENHAUSEN
in transmission electron microscopy
and
PETER ANDREWS, PH.D.
in scanning electron microscopy

Fourth Edition

77 Illustrations

Lea & Febiger

Philadelphia, 1973

Library of Congress Cataloging in Publication Data

Porter, Keith R.
   Fine structure of cells and tissues.

   First–2d ed. published under title: An introduction to
the fine structure of cells and tissues.
   Includes bibliographies.
      1. Histology—Atlases. 2. Cytology—Atlases.
I. Bonneville, Mary A., joint author. II. Title.
[DNLM: 1. Cytology—Atlases. 2. Histology—Atlases.
3. Microscopy, Electron. QS517 P846i 1973]
QM557.P67 1973   591.8′022′2   73–11314
ISBN 0-8121-0430-7

Fourth Edition

Copyright © 1973 by Lea & Febiger

All Rights Reserved

First Edition, 1963

Second Edition, September, 1964

Reprinted, October, 1964

Reprinted, January, 1966

Third Edition, September, 1968

Reprinted December, 1968

Published in Great Britain
by Henry Kimpton, London

German Translation
Springer-Verlag, Berlin

Spanish Translation
El Ateneo, Buenos Aires

Italian Translation
Piccin Editore, Padova

Japanese Translation
Homeido Shoten, Tokyo, 1966

Library of Congress Catalog Card Number: 73–11314

Printed in the United States of America

Preface

This edition is a very limited revision of the previous one. It includes minor changes in the basic text plus some new references to the more recent original literature. The expanded bibliographies are in no sense exhaustive; they serve rather to give the reader an introduction to some of the accumulated information on topics and phenomena represented by the original plates. With these references as a starter the zealous student can explore the discoveries that have become part of the literature since the previous edition appeared five years ago.

The same five-year-period has witnessed the development of scanning electron microscopy as a new approach to the study of cell fine structure. Because this means of observation is especially valuable in the study of the surfaces of tissues and cells, we have considered it essential to include a sampling of scanning micrographs in this edition. The micrographs selected are those that we believe should prove of value to the student in achieving an understanding of the nature of a complex surface. It is obvious that, as teaching aids in histology, scanning micrographs are unsurpassed and eventually will be used extensively. Thus far scanning microscopy has had only preliminary application in the fields of anatomy and pathology, but enough has been done to convince one of its values and to assure it a place as a standard procedure in anatomic studies.

The scanning images included in this edition were taken in our laboratories in the Department of Molecular, Cellular and Developmental Biology in Boulder, Colorado. We are especially indebted to Peter Andrews who did much to develop the techniques used and to Robert McGrew who was very effective in keeping the equipment functioning.

Keith R. Porter
Mary A. Bonneville

Boulder, Colorado

iii

Preface to Third Edition

This book, a limited collection of micrographs and associated legends, is designed to give the student of histology and cell biology a compact account of the more significant information currently available on cell and tissue fine structure. In this edition we have expanded the collection to include additional cell and tissue types that are exciting for the student and the investigator alike. Moreover, the text of the second edition has been largely rewritten. This was done primarily because the rapid accumulation of information on functional-structural relationships has made obsolete some of the earlier narratives. The objective, then, in this edition has been to provide information at the level required by the more sophisticated student now studying cell biology.

As we have worked on this edition, we have become aware, as have others interested in histology, that the microscopic anatomy of cells and tissues emerging from electron microscopy is rapidly gaining independent status and is superseding much of the older information from light microscopy. The latter, valuable in its time, is being pushed aside by this relatively recent accumulation of knowledge; many of the old questions and problems of thirty years ago have been solved, and entirely new ones are now being investigated. In what manner this shift of emphasis will influence the design of textbooks in cell biology and histology is not a subject for this limited preface except to note that it stimulates one to experiment with various ways of presenting structural information. The textbooks that will evolve will doubtless be different from those written during the first fifty years of this century.

It is our hope, of course, that this book will prove valuable to teachers and students of this subject. They may recognize that the atlas is, in a way, analogous to the classical slide collection, so much a part of past and current education in histology. It may indeed come to be used as a companion to conventional light microscope study. It also represents, we think, a body of information that can be used as a base on which to construct a more advanced course of study for cell and tissue fine structure. For this reason, we encourage reading in the current literature by including after each legend a short bibliography of significant papers.

To a very large extent, the pictures reproduced in this atlas originated in the Laboratory for Cell Biology at Harvard University and have not been published elsewhere. The authors are pleased to acknowledge the assistance of Robert Dell and Rick Stafford in the preparation of the photographic reproductions. To Helen Lyman we owe thanks for the line drawings and the cover design. Pamela Pettingill, our secretary, has dealt patiently and superbly with numerous drafts. Supplementary micrographs of special interest have been generously provided by investigators in other laboratories, and acknowledgments to them are made in the text. Finally, we are indebted to several friends who have contributed criticism and advice, and most especially to Helen Padykula, Geraldine Gautier, and A. Kent Christensen.

Keith R. Porter
Mary A. Bonneville

Cambridge, Massachusetts

iv

Contents

List of Abbreviations and Symbols

$\underline{A}$	A Band	Lu	Lumen
a, b, c, d	Regions of Intercalated Disk	Ly	Lysosome
Ac	Acrosome	$\underline{M}$	M Line
Ar	Arteriole	M	Mitochondrion
AS	Alveolar Sac	Mb	Microbody (Peroxisome)
BB	Basal Body	MD	Mucous Droplet
BC	Bile Canaliculus	Mt	Microtubule
BM	Basement Membrane (Basal Lamina)	Mv	Microvillus
Br	Bronchiole	My	Myelin
C	Cilium	N	Nucleus
Ca	Canaliculus	NE	Nerve Ending
Ce	Centriole	Nf	Neurofilaments
CeS	Centriolar Satellite	NF	Nerve Fiber
Ch	Chromatin	Ng	Neuroglial Cell
Chr	Chromosome	Ni	Nissl Body
Co	Collagen	Nu	Nucleolus
Col	Collecting Tubule	NuE	Nuclear Envelope
Cor	Stratum Corneum	O	Oocyte
Cp	Capillary	OR	Olfactory Rod
Cr	Crista	OV	Olfactory Vesicle
CT	Connective Tissue	P	Pore
D	Desmosome (Macula Adhaerens)	PC	Pigment Cell
DB	Dense Body	PCT	Proximal Convoluted Tubule
DT	Distal Tubule	PM	Plasma Membrane
E	Erythrocyte	Pr	Process
El	Elastic Fibers	Pt	Pit
Ell	Ellipsoid	R	Ribosome
En	Endothelium	RB	Residual Body
Ep	Epithelium	SC	Schwann Cell
ER	Endoplasmic Reticulum	SD	Secretion Droplet
F	Fibrocyte or Fibroblast	SER	Smooth Endoplasmic Reticulum
Fe	Ferritin	Sl	Slit
Fl	Filament	SM	Smooth Muscle
FP	Foot Process	Sn	Sinusoid
FS	Fibrous Sheath	Sp	Stratum Spinosum
FV	Fusiform Vesicle	SR	Sarcoplasmic Reticulum
FZ	Fibrillar Zone	St	Stalk
G	Golgi Complex	T	Tonofilaments and Tonofibrils
Ge	Stratum Germinativum	TB	Terminal Bar (Zonula Adhaerens)
Gl	Glycogen	Tc	Thrombocyte or Platelet
Gr	Granule	Th	Thymocyte
Gra	Stratum Granulosum	TS	Tubular System
Gu	Gutter	Tu	Tubule
GZ	Granular Zone	TW	Terminal Web
$\underline{H}$	H Band	UM	Unit Membrane
$\underline{\overline{H}}$	Hemoglobin	US	Urinary Space
HD	Hemosiderin Deposits	V	Vesicle
$\underline{I}$	I Band	$\underline{Z}$	Z Line
JF	Junctional Fold	$\overline{Z}$	Zymogen Granule
L	Lipid	ZA	Zonula Adhaerens (Terminal Bar)
Lc	Lymphocyte	ZO	Zonula Occludens
LF	Liquor Folliculi	ZP	Zona Pellucida
LH	Loop of Henle		

FINE STRUCTURE
OF CELLS
AND TISSUES

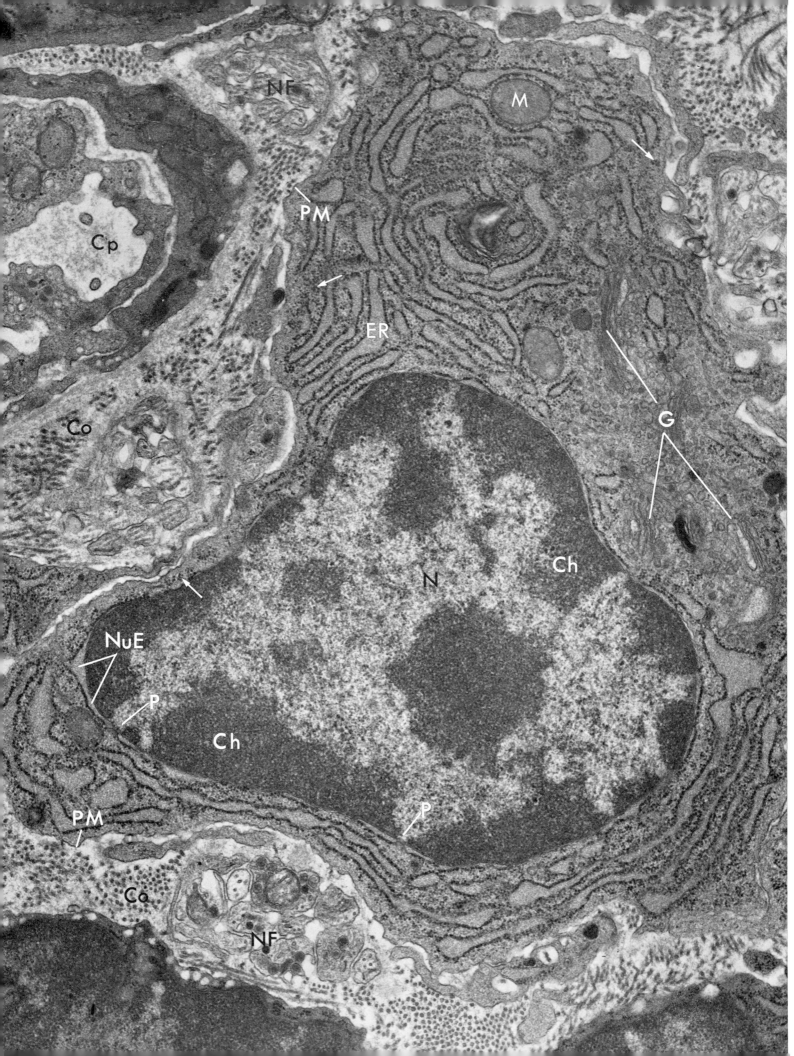

PLATE 1

The Plasma Cell

This micrograph of a plasma cell illustrates some of the structural detail revealed by electron microscopy and serves to introduce several of the systems and organelles repeatedly observed in most cell types. The clarity and richness of detail, especially of cytoplasmic structures, when observed even at relatively low magnifications, are a product of the resolving power of the electron optical system.

The plasma cell is readily identified under the light microscope by a combination of almost unmistakable features that are also easily recognized in electron micrographs. Following most fixation procedures (in this case glutaraldehyde followed by osmium[1]) the nucleus (N) takes on a "cartwheel" appearance due to the prominent clumping of dense granular chromatin (Ch) at its center and around its periphery. The chromatin, named because of its affinity for certain dyes, is the nuclear component richest in deoxyribonucleic acid or DNA. This is, of course, the material within which is coded the information necessary for maintenance of the cell's life and the determination of its special functions. In this dense form the chromatin is referred to by cytologists as heterochromatin and is believed to be in a relatively inactive state (see Plate 4). We might presume that information in the heterochromatin, while perhaps not functional in the plasma cell, had previously been decoded and used in some earlier cell generation. The less dense nuclear material, which displays little affinity for dyes in light microscope preparations, is called euchromatin. It is now thought that only in such regions is genetic information being transcribed for a particular cell. Note that the zones of euchromatin extend to the pores (P) in the nuclear envelope (NuE), while the heterochromatin is confined to the interpore areas. To reach the cytoplasm, information in the form of messenger RNA (one type of ribonucleic acid) must traverse the nuclear envelope, probably in the pore regions. (The nuclear envelope is discussed more fully in Plate 2, and details of nuclear structure are given in Plate 4.)

The appearance of the cytoplasm suggests to the experienced cell biologist that intense synthetic activity is occurring, since two systems known to be associated with the production of protein

for export from the cell, the rough-surfaced endoplasmic reticulum (ER) and the Golgi region (G), are both extensively developed. The markedly basophilic regions of plasma cell cytoplasm that characterize its light microscope image are now known with certainty (see Plate 2) to correspond to piles of large flattened sacs (cisternae) studded on their outer surfaces with ribosomes, particles of ribonucleoprotein. It is these that give the cytoplasm its special affinity for basic dyes. The cisternae shown here represent one structural form adopted by the endoplasmic reticulum, a membranous intracellular system that serves somewhat different functions in different cell types (see, for example, Plates 7, 11, 12, 14, 15, 18, 19, 38). The ribosomes are the sites of protein synthesis. In the plasma cell much of the protein product formed is sequestered within the cisternae and is seen in this micrograph as finely particulate material of medium density present within the sacs (see below).

The second cytoplasmic organelle of special prominence in the plasma cell is, as just mentioned, the Golgi complex (G). In the plasma cell it occupies an extensive juxtanuclear region and displays its typical configuration, i.e., stacks of flattened sacs and associated vesicles. Both are delimited by smooth-surfaced membranes. The Golgi, which appears under the light microscope as an unstained area when basic dyes are applied, was for many years the subject of controversy and discussion among light microscopists, many of whom regarded it as an artifact. At present not only is its morphology well established, but its several functions are being revealed. Prominent among these is the function of packaging protein and complex carbohydrate products of the cell into membrane-bounded granules that are subsequently secreted (see Plates 11 and 12).

The mitochondria (M), although not particularly numerous in plasma cells, are easily identified by their size and general structural plan (described in connection with Plate 2). They are well known as organelles active in the production of energy-rich compounds by aerobic metabolic pathways.

The organelles described above are embedded in a cytoplasmic ground substance. Among its several functions it serves as a reservoir for compounds that will be needed to produce secretory products (the major activity of the plasma cell) and

[1] See page 197 for general description of techniques.

to repair the structural components of the organelles themselves. Free ribosomes (arrows) are included in the ground substance and in this location are thought to participate in the formation of proteins that are retained within the cell as part of the cytoplasmic machinery.

Plasma cells—and indeed all cells—are enclosed in a plasma membrane (PM), which in transverse sections appears as a thin dense line. This lipoprotein structure is semipermeable; that is, it acts as a selective barrier, exerting control over the passage of ions and small molecules into and out of the cell. Certain molecules are prevented from entering, but others are taken in or expelled preferentially by "active" transport. The latter term implies that energy is expended to overcome adverse concentration gradients and to "pump" essential substances across the membrane. The basic substructure common to plasma membranes and to cytoplasmic membranes is discussed in Plate 2.

The highly ordered and complex cytoplasmic organization of the plasma cell is a manifestation of its mature or differentiated state, in which it performs special functions. Several lines of evidence indicate that this cell, which is found in certain areas of connective tissue, is most important in the formation of circulating antibodies. By this, of course, is meant that in response to a number of foreign materials (called antigens) these cells produce proteins (antibodies), which combine specifically with the foreign antigens. The animal containing the responding plasma cells is then said to be immunized. This highly specific reaction serves to protect the organism against the harmful effects of extraneous substances.

Direct evidence of plasma cell involvement in the synthesis of antibodies and information on the precise localization of the sites of synthesis within the cell can now be obtained by electron microscopy. If, for example, the plasma cells from an animal immunized against horse ferritin (an iron-rich protein) are appropriately fixed, exposed to ferritin, and then prepared for electron microscopy, it is found, as illustrated in the accompanying text figure 1a, that the ferritin

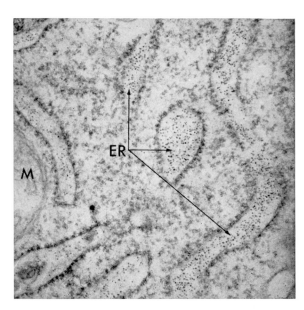

Text Figure 1a

This electron micrograph, kindly supplied by S. de Petris, shows a part of a plasma cell from a rabbit hyperimmunized against ferritin. This cell and others were briefly fixed with formalin, frozen (to open the cells), and thawed in the presence of antigen. The dense particles in the ER cisternae (ER) represent ferritin in combination with antibody. A mitochondrion (M) is included in part at the left.

Magnification × 61,000

molecules can be visualized as dense particles (because of their iron cores) and that they are associated specifically with the contents of the ER cisternae. This reaction marks the location of the antibody and shows that, like other proteins, it is segregated in these cavities after synthesis.

4

Similar observations have been made using other proteins as antigens and then linking these to ferritin as a marker. These experiments stem in turn from earlier fundamental studies in which antigen or antigen-antibody complexes were localized in tissues with the aid of fluorescent dyes and light microscopy.

The plasma cell in this micrograph lies within a connective tissue ground substance rich in collagen fibers (Co). Bundles of thin unmyelinated nerve fibers (NF) and the cross section of a small blood vessel (Cp) may also be identified.

In the plasma cell and in other cells and tissues presented in this atlas are found homologous structures that occur almost universally in living cells; these are the structures concerned with maintaining processes vital to the life of the individual cell. In addition, these same components show great diversity and specialization related to the performance of specific functions, i.e., functions peculiar to each of the wide variety of cell and tissue types characteristic of multicellular organisms. The correlation of this diversity in form at the fine structural level with diversity in function, which is the current interest of many investigators, makes this period one of the more exciting in the history of cell biology.

From the submucosa of the rat (*Mus norvegicus*)
Magnification × 29,000

5

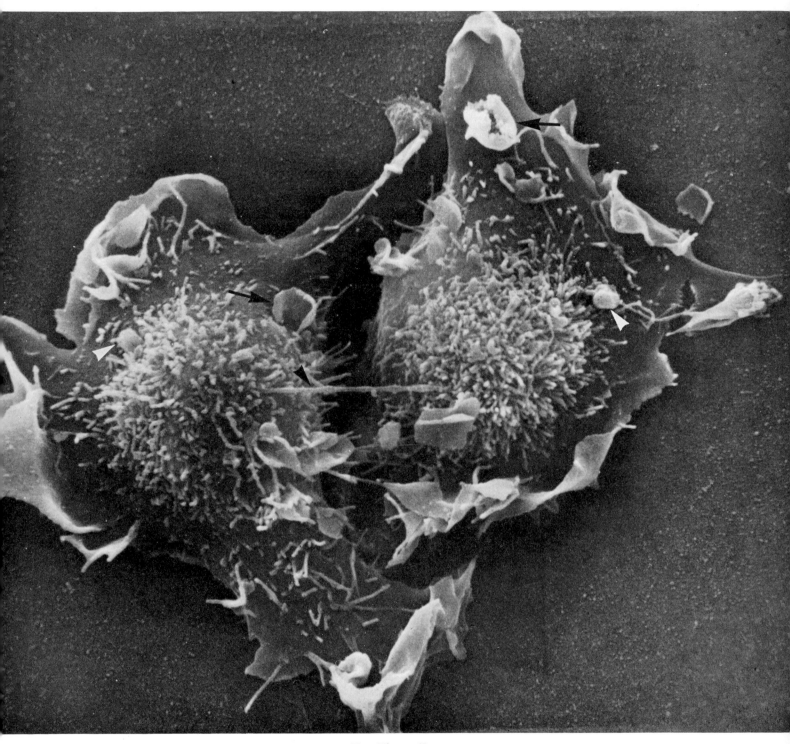

Text Figure 1b
Surface Features of Cultured Cells

6

Surface Features of Cultured Cells (Text Figure 1b)

The surfaces of cells show a wide range of variation in both form and function. Some, such as the free surfaces of adsorptive cells of the intestine (Plates 6 and 7) and the proximal convoluted tubule cells of the kidney (Plate 24), display numerous microvilli. Others are ciliated, like the cells of the tracheal epithelium (Plate 8), while still others are without evident differentiations or specializations (Plate 33). There are probably as many different surfaces as there are types of differentiated cells in the vertebrate organism.

Their differences notwithstanding, it is clear that all surfaces have a few features in common. For example, each surface regulates the flow of metabolites into and out of the cell; each surface is covered with characteristic glycoproteins that influence the capacity of cells to recognize others of the same kind; and, by means of their surfaces, contacting cells may inhibit the growth, movement, and behavior of one another. The variation in important functions and related structures makes examination of cell surfaces in as many ways as possible attractive for the investigator. Until very recentlly, direct observation of the morphologic characteristics of these surfaces was limited to views of thin sections or of surface replicas that were provided by electron microscopy. Although this approach has its values, as well as limitations, it does not provide images of complete surfaces. At best, with this method, complete images have to be constructed from fragments of information.

Fortunately the opportunities for direct viewing of surfaces are greatly improved by the scanning electron microscope. In image formation, this new instrument makes use of secondary electrons emitted from the specimen's surface while it is being scanned by a small pencil of high-energy electrons. The result is an image that defines the contours of the surface being scanned, has a depth of focus amounting to several micrometers, and shows a resolution at least tenfold better than that provided by the light microscope.

The techniques of specimen preparation are quite simple. The cells or soft tissues to be examined are fixed in glutaraldehyde and osmium tetroxide, essentially as for embedding and thin sectioning. They are then washed in water, dehydrated in acetone, and transferred to a pressure chamber (a bomb) where liquid carbon dioxide is substituted for acetone. When the temperature of the bomb is raised, the liquid carbon dioxide is brought to its critical point (32° C), at which it changes to a gas without change of volume and without the development of any liquid-gas interphase. As a consequence the specimen is never exposed to the forces of surface tension while drying. The pressure in the bomb is slowly released and the dry specimen is essentially ready for examination. A coating of gold or gold-palladium, applied in a final step by evaporation in a vacuum chamber, gives the surface of the specimen the required conductivity as well as image-forming properties.

The two cells shown in the scanning micrograph (figure 1b) are part of a population grown *in vitro* from explants of a rat sarcoma. Therefore they are tumor cells and should not be regarded as normal in their surface properties. Nevertheless they do not show any type of surface structure that is not also found on normal cells.

Since the cells are still connected by a slender bridge (black arrowhead) we know that they have only recently divided (possibly two to three hours earlier) and are in early G_1 of the cell cycle. It is obvious that they are beginning to spread out over the culture substrate, a process that may continue through S and into G_2.

These particular cells show two major surface excrescences. There are small fingerlike projections clustered over the elevated centers of the cells. These are uniform in diameter (about 0.1 μm), but vary in length from a fraction of 1 μm up to several micrometers. Though not easily observed in the living cell by any available device, the formation and regression of these structures probably goes on continuously. While their function is not known in this instance, they should, in enhancing the total surface area, facilitate the exchange of metabolites between the cell and its environment.

Also these cells have ruffles—the thin lamellar (sail-like) vertical extensions of the surface that are most evident at the cell margins. The ruffles are only 0.1 to 0.2 μm thick, but may have surface areas of a hundred or more square micrometers. In the living cell they seem to form at the cell margins in only a few minutes and then move centrally over the surface. In the course of this migration the individual ruffle seems to form a small cup (single arrow) which then closes over to engulf a small quantity of the surrounding medium. This is the structural basis for pinocytosis (cell drinking).

Other surface structures occasionally seen include zeiotic blebs and filopodia. Neither of these is displayed to good advantage here.

Blebs are small spherical extensions of the surface (white arrowhead). They may also vary tremendously in size from approximately 0.2 to 5.0 μm. Although common during the late phases of mitosis and during early G_1, they tend to be absent during S and G_2. Their function is unknown and it helps little to learn from thin sectioning, as one can, that they are packed with ribosomes.

Filopodia are not easily separated from microvilli. They are ordinarily very long and more slender than microvilli and are seemingly involved in connecting cell to cell and cell to substrate.

These are only a fraction of the surface features that may eventually be revealed by scanning electron microscopy. Others include enzymes and antigenic substances or binding sites for hemagglutinins—all too small for visualization by this mode of scanning. If, however, the antibody or the hemagglutinin is labeled with a large molecule (e.g., hemacyanin, 300 A), it can serve as a marker for the location of the antigen or the binding site. These and other techniques are under development and promise to open up a whole new approach to cell surface investigations.

Rat sarcoma cell in culture
Magnification × 4,000

Note: The first five plates of this edition of the atlas serve as a general introduction to the concepts of cell structure in accord with current knowledge, especially that derived from electron microscopy. Descriptions of cell organelles touched on briefly in Plate 1 are referred to in greater detail in Plates 2-5, where pertinent references to each are given. The student may find it helpful to read introductory chapters in standard textbooks of histology before studying original research papers. The article by Brachet, cited below, will also be helpful. The remaining references given here deal with the specific function of the plasma cell and with scanning electron microscopy.

ABERCROMBIE, M., HEAYSMAN, J. E. M., and PEGRUM, S. M. The locomotion of fibroblasts in culture. II. "Ruffling." *Exp. Cell Res., 60*:437 (1970).

ANDERSON, T. F. Techniques for the preservation of three-dimensional structure in preparing specimens for the electron microscope. *Trans. N.Y. Acad. Sci., Ser. II, 13*:130 (1951).

AVRAMEAS, S., and LEDUC, E. H. Detection of simultaneous antibody synthesis in plasma cells and specialized lymphocytes in rabbit lymph nodes. *J. Exp. Med., 131*:1137 (1970).

BOYDE, A., WEISS, R. A., and VESELY, P. Scanning electron microscopy of cells in culture. *Exp. Cell Res., 71*:313 (1972).

BRACHET, J. The living cell. *Sci. Amer., 205*:51 (September, 1961).

COONS, A. H., LEDUC, E. H., and CONNOLLY, J. M. Studies on antibody production. I. A method for the histochemical demonstration of specific antibody and its application to a study of the hyperimmune rabbit. *J. Exp. Med., 102*:49 (1955).

COSTERO, I., and POMERAT, C. M. Cultivation of neurons from the adult human cerebral and cerebellar cortex. *Amer. J. Anat., 89*:405 (1951).

DE PETRIS, S., KARLSBAD, G., and PERNIS, B. Localization of antibodies in plasma cells by electron microscopy. *J. Exp. Med., 117*:849 (1963).

EVERHART, T. E. and HAYES, T. L. The scanning electron microscope. *Sci. Amer., 226*:54 (January, 1972).

FOLLETT, E. A. C., and GOLDMAN, R. D. The occurrence of microvilli during spreading and growth of BHK21/C13 fibroblasts. *Exp. Cell Res., 59*:124 (1970).

NEMANIC, M. K., and PITELKA, D. R. A scanning electron microscope study of the lactating mammary gland. *J. Cell Biol., 48*:410 (1971).

PORTER, K. R., KELLEY, D., and ANDREWS, P. M. The preparation of cultured cells and soft tissues for scanning electron microscopy. *In* Proceedings, 5th Annual Stereoscan Scanning Electron Microscope Colloquium. Morton Grove, Illinois, Kent Cambridge Scientific Company (1972) p. 1.

————, PRESCOTT, D., and FRYE, F. Changes in surface morphology of CHO cells during the cell cycle. *J. Cell Biol., 57*:815 (1973).

PRICE, Z. H. The micromorphology of zeiotic blebs in cultured human epithelial (HEp) cells. *Exp. Cell Res., 48*:82 (1967).

RIFKIND, R. A., OSSERMAN, E. F., HSU, K. C., and MORGAN, C. The intracellular distribution of gamma globulin in a mouse plasma cell tumor (X5563) as revealed by fluorescence and electron microscopy. *J. Exp. Med., 116*:423 (1962).

SCHUTZ, L., and MORA, P. T. The need for direct cell contact in "contact" inhibition of cell division in culture. *J. Cell. Physiol., 71*:1 (1968).

PLATE 2

Cell Systems and Organelles:
Endoplasmic Reticulum, Nuclear Envelope, Plasma Membrane, and Mitochondria

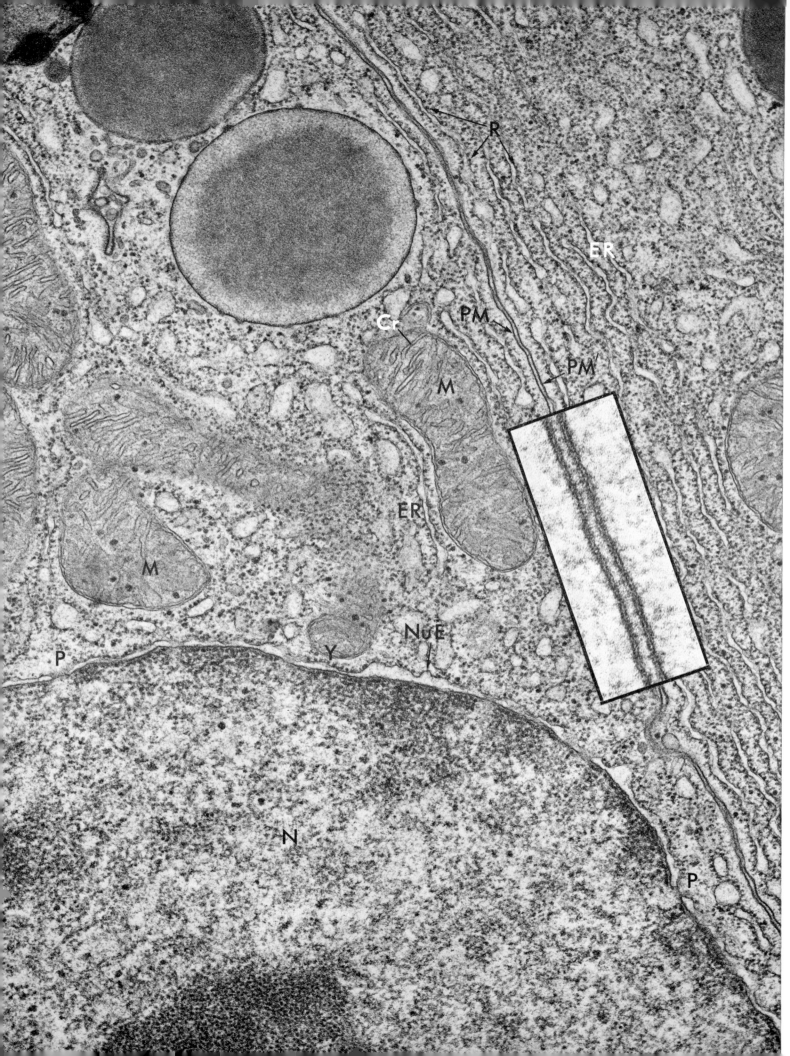

PLATE 2

Cell Systems and Organelles: Endoplasmic Reticulum, Nuclear Envelope, Plasma Membrane, and Mitochondria

From viewing this and the preceding plate one may reasonably gain the impression that the cytoplasm of cells is a confusion of membranes and particles so complex as to defy understanding. Though quite justified, this first impression has not apparently discouraged attempts to unravel the obvious complexities; instead, it has probably served to stimulate their investigation. Thus some cell biologists have undertaken widely ranging comparative studies of different cell types to determine how commonly the observed structures occur and to note any constant correlation between form and known functional specialization. Others have essentially disassembled the cell and isolated by centrifugation fractions rich in one or the other of the cytoplasmic substructures. Straightforward analyses of these fractions for their composition and enzymic properties have given valuable information as to the intracellular distribution of biochemical processes and clues to the significance of various structural features. It is now realized, for example, that membranes, which occur in such great profusion in some cytoplasmic systems and organelles, provide relatively large surfaces upon which packets of enzymes and genetic information may be distributed according to a functionally important plan. The spaces limited by such membranes serve as sites for the sequestration of products of the cell's synthetic activity. The organized and frequently patterned distribution of finely divided substructures and surfaces seems designed to bring enzyme and source of substrate closer together or energy-rich compound closer to where it is needed. Diffusion distances, which could be rate limiting for physiological processes, are thus made less significant.

One of the more remarkable of the systems involved in these several roles is called the endoplasmic reticulum (ER). This complex of membrane-limited spaces extends to nearly all parts of the cytoplasm and thus endows the cell with an extensive intracytoplasmic compartment, separate from the cytoplasmic ground substance. In a thin section of a fully differentiated cell of the pancreas, shown here and also in the plasma cell shown in Plate 1, the ER is constructed in large part of thin flattened sacs, or cisternae, which appear in section profile as long, slender, membrane-limited spaces of homogeneous content suspended in a particle-rich matrix. Many of the 200 to 250 A particles, the ribosomes (R), are attached to the outer surfaces of the cisternae—an association that is universal in cells manufacturing proteins for export. Visualization of the entire system as it exists in a cell is, of course, impossible from a single thin section, but from the study of many sections cut at different angles and planes through the cell the total form can be reconstructed. Actually the existence of this system, as well as its spatial arrangement, was first observed in electron micrographs of thinly spread whole cells, where it could be seen *in toto*.

Our concepts of the endoplasmic reticulum, which have greatly expanded in recent years, are actually based on observations going back to the beginning of the century. In 1899, Garnier and Matthews independently described the basophilic properties of certain regions of the cytoplasm, called the ergastoplasm, and pointed out that this basophilic component is prominently developed in certain secretory cells. Several decades passed before the distribution of this component in the cytoplasm was shown to coincide with the distribution of ribonucleic acid (RNA). This discovery rested mainly on two observations, namely that the property of basophilia could be destroyed by the enzyme ribonuclease (which hydrolyzes RNA) and that the ergastoplasm, like RNA, strongly absorbs ultraviolet light in the wavelength of about 260 nm.

Another exciting phase of this work began in the late 1940's, when correlative experiments involving electron microscopy and differential centrifugation resolved, with a wealth of detail formerly impossible, the structure and function of the ergastoplasm. A tiny vesicular element, the microsome of cell fractions, was identified as a fragment of the rough-surfaced endoplasmic reticulum. And labeled amino acids, provided experimentally to an animal's circulation, quickly became part of proteins in the microsomes. For the first time, then, the investigator could identify the rough-surfaced endoplasmic reticulum with protein synthesis. Experiments of a similar kind later narrowed this function to the RNA-rich ribosomes. Further reference to this provocative work is given in the text for Plate 11.

Little concrete information is available concerning the function of the ER membranes. When high resolution micrographs are obtained, these membranes can be shown to share a common

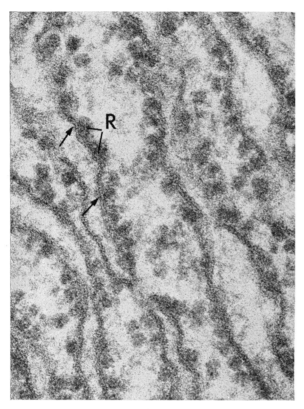

Text Figure 2a

In this high resolution electron micrograph, the ribosomes (R) and their relation to the ER membranes are particularly well shown. Like other cytoplasmic membranes, those of the ER are trilaminar (arrows). The ribosomes are closely adherent to the membrane, but whether at the zone of contact they are structurally integrated with the outer leaflet of the membrane cannot be easily decided. It is pertinent in this regard that the attachment is not a permanent one, inasmuch as the ribosomes are known to dissociate from the membrane surface. (This micrograph has been kindly provided by E. De Harven and is reproduced from *Methods in Cancer Research*, vol. I, H. Busch, editor, New York, Academic Press, 1967.)

From the liver of the mouse (*Mus musculus*)
Magnification × 150,000

structure with other cytoplasmic membranes (see below and also text figure 2a). That they have special permeabilities is manifest in their ability to sequester within the cavities they limit products synthesized by the ribosomes on their surfaces. Furthermore, in the living cell the membranes seem extraordinarily labile and lend themselves to the remodeling that the ER constantly undergoes in response to shifts in physiological conditions. The remodeling may involve changes in area, modification from rough to smooth form and from lamellar cisternae to reticular, as well as changes in location relative to the mitochondria, the Golgi, and the cell surface.

Thin sections of the kind depicted here also reveal a close relationship between the ER and the membrane system surrounding the nucleus (the nuclear membrane of classical cytology). As can be seen, the nucleus (N) is enclosed by a system of two membranes, aptly named the nuclear envelope (NuE). The space enclosed by these membranes has been repeatedly observed to be continuous with that within the membranes of the endoplasmic reticulum. As a consequence the envelope can be regarded as a special portion of the ER, or, if one prefers, the cytoplasmic ER is an outgrowth of the envelope that is ever present in nucleated cells. Its outer surface at least is like ER membranes, studded with ribosomes (as at Y); its inner membrane is in close contact with the heterochromatin (see Plates 1 and 4).

The nuclear envelope has a few other interesting properties. It is, for example, interrupted by pores (P). Moreover, the inner and outer membranes are continuous around the rim of the pore. In spite of this arrangement it seems unlikely that there is free passage for large molecules between the interphase nucleus and the cytoplasm of most cell types, for in micrographs of excellent quality a diaphragm of dense material is often found filling each gap in the envelope. At least one investigator has shown that this constitutes a barrier for macromolecules like ferritin but not for colloidal gold. These observations have aroused interest because the currently accepted theory of protein synthesis requires movement of RNA molecules (e.g., messenger, transfer, and ribosomal RNA's) between nucleus and cytoplasm.

Different sorts of membranes are distinguishable in electron micrographs. For instance, the plasma membranes (PM) and (PM') of the two adjacent secretory cells differ from the cytoplasmic membranes in several respects: they are with-

12

out attached ribosomes, and they are slightly thicker (100 A) than those of the ER (70 A). Except at isolated and special sites, the surface membrane of one cell never fuses with its neighbor's. Instead, it is usual for the plasma membrane of adjacent cells to be separated by a distance of at least 200 A, a fact taken to indicate the presence of a thin layer of material (a mucopolysaccharide-rich layer called the glycocalyx) on the outer surface of each membrane. As apparent in the inset on this figure, each plasma membrane can be resolved at high magnifications into two dense layers and an intermediate layer of lower density. This trilaminar structure has been called the unit membrane since it seems to reflect a universally occurring molecular organization in plasma membranes and cytoplasmic membranes alike (see text figure 2a). The staining pattern observed in electron micrographs results from the interaction of osmium tetroxide with the phospholipid bilayer forming the backbone of biological membranes.

Maintaining complex cell structure and carrying out various physiological processes impose a constant need for energy on living cells. Living systems as we know them utilize as an immediate source of energy the compound adenosine triphosphate (ATP), which they are able to manufacture and to some degree store. Both aerobic and nonaerobic (anaerobic) processes can produce ATP, but in the cells of vertebrates the former is the more efficient and important. By again exploiting the techniques of cell fractionation, biologists showed some twenty years ago that cytoplasmic particles with certain sedimentation properties were capable of oxidizing fatty acids and producing ATP aerobically. Microscopic examination of the granules established that they were mitochondria, structures long recognized as an almost universal component of cytoplasm.

Once techniques for thin sectioning of cells were developed, it was found that mitochondria (M) possess a common structural plan. Their identification in sections is thus assured. Variations in their morphology exist from cell type to cell

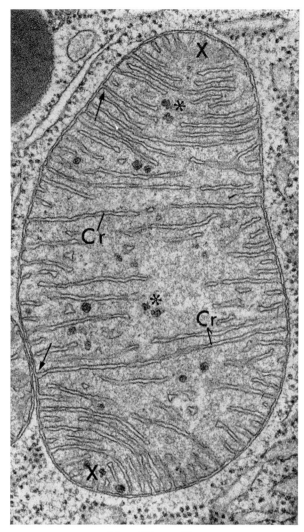

Text Figure 2b

A longitudinal section of a mitochondrion from the pancreas of the bat exemplifies the basic structural plan common to all these organelles. The arrows indicate places at which the inner of the two membranes limiting the mitochondrion is clearly continuous with that forming the cristae (Cr). Dense granules (*) are suspended in a relatively less dense matrix. Mitochondria are known to grow in length, and growing points of this organelle are perhaps represented by areas (X) at either end of it.

From the pancreas of the bat
Magnification × 64,000

type, but they are always seen to be limited by two lines that represent the two membranes constituting the wall of these tiny organelles. As revealed at the higher magnification shown in text figure 2b, the inner of the two membranes is folded into shelf-like structures, the cristae (Cr), which project into the homogeneous matrix that fills the cavity of the mitochondrion. Where the energy requirements of the cell are greater, as in heart muscle (Plate 39), the mitochondria are more numerous, and the number of cristae is greater, presumably to provide a larger surface area for the organized distribution of the resident enzymes. Dense granules, 20–30 nm in diameter, also evident in the matrix, are now known to represent accumulations of bound divalent metallic ions (Ca^{++} and Mg^{++}) required for mitochondrial enzyme systems.

Electron micrographs of course present a completely static picture, but in time-lapse cinematography of living cells it is strikingly apparent that mitochondria stretch and contract, move about in the cell, and seem to coalesce and divide. Such properties as growth and division suggest a measure of autonomy or semi-autonomy for mitochondria, and a number of recent studies of them are consistent with this notion. For example, contractile protein, as well as specific nucleic acids (both RNA and DNA), has been found within them. There has also been advanced more sophisticated evidence that new mitochondria arise only by division of pre-existing mitochondria.

From the pancreas of the frog (*Rana pipiens pipiens*)
Magnification × 43,000
Inset × 216,000

References

ADELMAN, M. R., SABATINI, D. D., and BLOBEL, G. Ribosome-membrane interaction. Nondestructive disassembly of rat liver rough microsomes into ribosomal and membranous components. *J. Cell Biol., 56*:206 (1973).

BENNETT, H. S. Morphological aspects of extracellular polysaccharides. *J. Histochem. Cytochem., 11*:14 (1963).

CLAUDE, A. Fractionation of mammalian liver cells by differential centrifugation. I. Problems, methods, and preparation of extract. *J. Exp. Med., 84*:51 (1946).

———. Fractionation of mammalian liver cells by differential centrifugation. II. Experimental procedures and results. *J. Exp. Med., 84*:61 (1946).

DE PIERRE, J. W., and KARNOVSKY, M. L. Plasma membranes of mammalian cells. A review of methods for their characterization and isolation. *J. Cell Biol., 56*: 275 (1973).

FELDHERR, C. M. Binding within the nuclear annuli and its possible effect on nucleocytoplasmic exchanges. *J. Cell Biol., 20*:188 (1964).

FOX, C. F. The structure of cell membranes. *Sci. Amer., 226*:30 (February, 1972).

LEHNINGER, A. L. The Mitochondrion. Molecular Basis of Structure and Function. New York, W. A. Benjamin, Inc. (1964).

LITTLEFIELD, J. W., KELLER, E. B., GROSS, J., and ZAMECNIK, P. C. Studies on cytoplasmic ribonucleoprotein particles from the liver of the rat. *J. Biol. Chem., 217*:111 (1955).

LUCK, D. J. L. Formation of mitochondria in *Neurospora crassa*. A quantitative radioautographic study. *J. Cell Biol., 16*:483 (1963).

MARCHESI, V. T., TILLACK, T. W., JACKSON, R. L., SEGREST, J. P., and SCOTT, R. E. Chemical characterization and surface orientation of the major glycoprotein of the human erythrocyte membrane. *Proc. Nat. Acad. Sci. U.S.A., 69*:1445 (1972).

NOMURA, M. Ribosomes. *Sci. Amer., 221*:28 (October, 1969).

PALADE, G. E. A small particulate component of the cytoplasm. *J. Biophysic. and Biochem. Cytol., 1*:59 (1955).

———. An electron microscope study of the mitochondrial structure. *J. Histochem. Cytochem., 1*:188 (1953).

———, and SIEKEVITZ, P. Liver microsomes: an integrated morphological and biochemical study. *J. Biophysic. and Biochem. Cytol., 2*:171 (1956).

PIKÓ, L., BLAIR, D. G., TYLER, A., and VINOGRAD, J. Cytoplasmic DNA in the unfertilized sea urchin egg: physical properties of circular mitochondrial DNA and the occurrence of catenated forms. *Proc. Nat. Acad. Sci. U.S.A., 59*:838 (1968).

PORTER, K. R. The endoplasmic reticulum: some current interpretations of its forms and functions. *In* Biological Structure and Function. T. W. Goodwin and O. Lindberg, editors. New York, Academic Press, vol. I (1961) p. 127.

———. Observations on a submicroscopic basophilic component of cytoplasm. *J. Exp. Med., 97*:727 (1953).

REDMAN, C. M., and SABATINI, D. D. Vectorial discharge of peptides released by puromycin from attached ribosomes. *Proc. Nat. Acad. Sci. U.S.A., 59*: 608 (1966).

———, SIEKEVITZ, P., and PALADE, G. E. Synthesis and transfer of amylase in pigeon pancreatic microsomes. *J. Biol. Chem., 241*:1150 (1966).

REICH, E., and LUCK, D. J. L. Replication and inheritance of mitochondrial DNA. *Proc. Nat. Acad. Sci. U.S.A., 5*:1600 (1966).

RIFKIN, M. R., WOOD, D. D., and LUCK, D. J. L. Ribosomal RNA and ribosomes from mitochondria of *Neurospora crassa. Proc. Nat. Acad. Sci. U.S.A., 58*: 1025 (1967).

ROBERTSON, J. D. Design principles of the unit membrane. *In* Principles of Biomolecular Organization. Ciba Foundation Symposium, G. E. W. Wolstenholme and M. O'Connor, editors. London, J. and A. Churchill Ltd. (1966) p. 357.

SABATINI, D. D., TASHIRO, Y., and PALADE, G. E. On the attachment of ribosomes to microsomal membranes. *J. Molec. Biol., 19:*503 (1966).

SIEKEVITZ, P., and PALADE, G. E. A cytochemical study on the pancreas of the guinea pig. V. *In vivo* incorporation of leucine-1-C^{14} into the chymotrypsinogen of various cell fractions. *J. Biophysic. and Biochem. Cytol., 7:*619 (1960).

SINGER, S. J., and NICOLSON, G. L. The fluid mosaic model of the structure of cell membranes. *Science, 175:*720 (1972).

SJÖSTRAND, F. S., and ELFVIN, L.-G. The layered, asymmetric structure of the plasma membrane in the exocrine pancreas cells of the cat. *J. Ultrastruct. Res., 7:*504 (1962).

SLAYTER, H. S., WARNER, J. R., RICH, A., and HALL, C. E. The visualization of polyribosomal structure. *J. Mol. Biol., 7:*652 (1963).

STOECKENIUS, W. Structure of the plasma membrane. *Circulation, 26:*1066 (1962).

WARNER, J. R., KNOPF, P. M., and RICH A. A multiple ribosomal structure in protein synthesis. *Proc. Nat. Acad. Sci. U.S.A., 49:*122 (1963).

WATSON, M. L. Further observations on the nuclear envelope of the animal cell. *J. Biophysic. and Biochem. Cytol., 6:*147 (1959).

YAMAMOTO, T. On the thickness of the unit membrane. *J. Cell Biol., 17:*413 (1963).

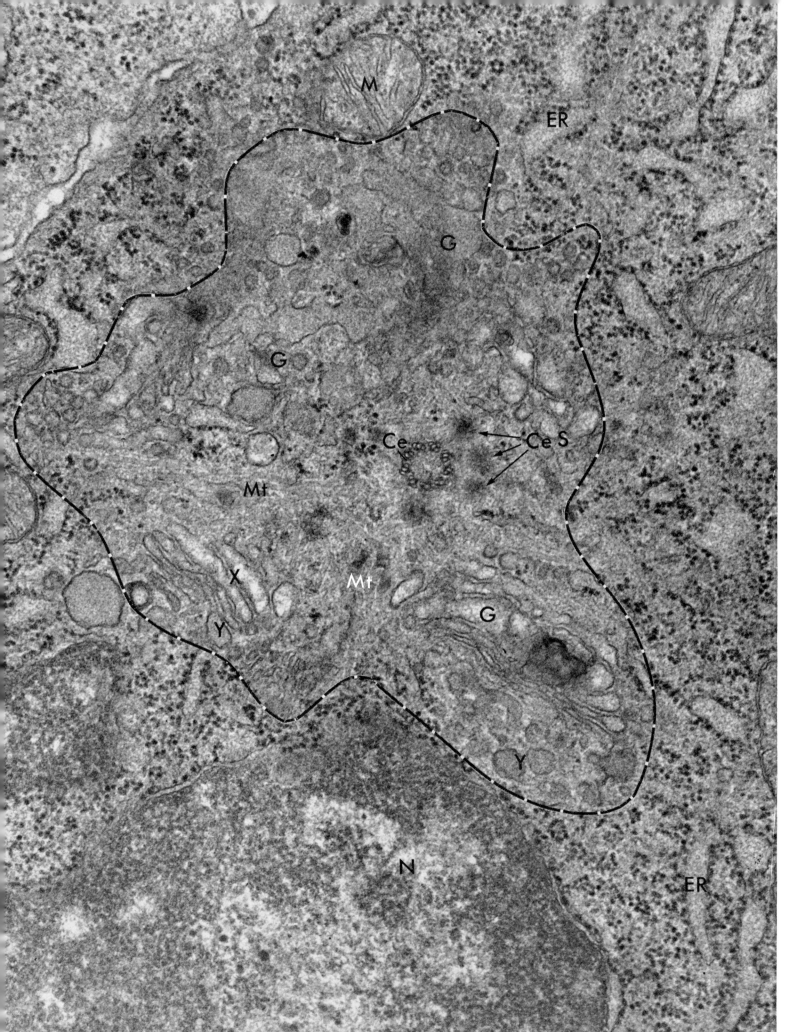

PLATE 3

Cell Systems and Organelles:
The Cell Center with Centriole and Golgi Complex

It is common for tissue cells to show evidence of polarity or a roughly radial organization around a center called the cytocentrum or centrosphere. In animal cells two paired structures of unique design, the centrioles, generally reside at the focal point of this organization. In order to discover the nature of the influence (structural or physiological) that emanates from this point, investigators must apply a variety of techniques in addition to electron microscopy. It must be admitted, however, that even now no completely clear understanding of the basis for cell polarity and organization has been achieved.

The cytocentrum, depicted here within the broken line, lies near one pole of the nucleus (N). Most of the larger cell organelles, such as the mitochondria (M) and rough-surfaced endoplasmic reticulum (ER), are excluded from it. Micromanipulation studies reveal the centrosphere to be more gelatinous than other parts of the cytoplasm, so that free movement of organelles into the region may be prevented. The central zone of this region is occupied by the paired cylindrical centrioles (0.2 μm in diameter), of which one is shown here in cross section (Ce), together with closely associated centriolar satellites (CeS). (The second centriole, if evident in the section, would be oriented with its long axis at 90° to that of the one shown.) A myriad of microtubules (Mt), quite closely packed, radiate from this central zone. These are slender (240 A diameter) rods, viewed here for the most part as running obliquely through the section.

In its fine structure, as shown here in cross section, the centriole is made up of nine triplets consisting of small, tubular structures (microtubules). The triplets are equidistant from one another and from the central axis of the bundle. It is evident as well that this precise organization extends to the small masses of dense material, the satellites, for they too are radially arranged at a uniform distance about the centriolar axis. Occasionally a second or even a third order of these bodies can be identified.

Near the margins of the cytocentrum in this plasma cell one observes elements of the Golgi complex (G). As mentioned in the text of Plate 1, this is a major intracellular system of membrane-limited sacs, which can now be easily identified in a wide variety of cell types and which frequently exists in the shape of a basket around the central pair of centrioles. Though not displayed here with diagrammatic clarity, some features of the Golgi morphology can be discerned. These will be found repeatedly in other displays of the complex included in this collection of micrographs (see Plates 11, 12, 17 and 23).

The aspect of the Golgi most easily recognized is that shown at X. It represents a section cut transversely through a stack of cisternae. These take the form of flattened sacs, which here number four to eight but vary in number within the Golgi complexes of different kinds of cells. They have associated with them the profiles of various smaller and larger rounded vesicles. The stacks of cisternal elements are usually curved and therefore have a concave and convex side. This configuration imparts a polarity to the stacks. In this micrograph vesicles enclosing material of low density are adjacent to the convex surface (as at X). Other vesicles on the concave side show more extensive size variation, with the larger ones (Y) containing a homogeneous substance of medium density.

The possible structural relationship between the membranous Golgi components has been deduced from clues gleaned from electron micrographs. It is certain that the system is a dynamic one, and images have been seen indicating that the small vesicles pinch off the margins of the flattened sacs. Once formed, the vesicles seem capable of growing or coalescing to form larger membrane-bounded bodies that may then move into other areas of the cytoplasm or even to the cell surface. It is also generally accepted that vesicular derivatives of ribosome-free surfaces of the endoplasmic reticulum are fed into the convex pole and integrated into the Golgi complex. Therefore, while a fixed connection between these two major membrane systems is not present, the contents and membranes of one may be transferred to the other.

The characteristics of the Golgi complex are sufficiently constant and well enough known to make certain its identification in electron micrographs. This is, however, a relatively recent capability. In fact, for several decades microscopists were not convinced even of the reality of this cell system. This uncertainty resulted from the apparent variability of form displayed by the Golgi and the limited resolving power of the light microscope. As originally described by Golgi (in 1898), it appeared as a reticular structure stained black by its ability to reduce silver or osmium to metallic form. In spite of careful light microscope studies, especially those of Bowen, on the formation of the acrosome in sperm (see Plate 23), unbelievers persisted. Finally evidence from electron microscopy provided convincing identification and a relatively complete description. The Golgi is now accepted as a major membrane-limited system of both plant and animal cells, and evidence is accumulating to indicate that, like several other cell components, it possesses the capacity to grow and duplicate.

For several decades those cytologists willing to accept the existence of the Golgi believed it to be related in some way to secretion. This role has now been investigated extensively, particularly in the exocrine cells of the pancreas (see Plate 11). In those cells, as well as in other protein-secreting cells, it is thought that specific cell products, synthesized by the ribosomes and sequestered within the membranes of the ER, are transferred into the Golgi spaces and there packaged into membrane-bounded secretory droplets, to be ejected later from the cell. The Golgi is also known to participate in the production of specific enzyme-containing vesicles that fuse with and contribute hydrolytic enzymes to lysosomes (see Plate 15). In addition it also brings about the sulfation of mucopolysaccharides within cells that produce mucus (Plate 12) and extracellular ground substance (Plates 28 and 29). It may therefore be regarded in general as a site of assembly and production of cell products that may be stored in membrane-bounded granules before their use within or secretion from the cell.

From a plasma cell in the intestinal submucosa of the rat
Magnification $\times$ 67,000

References

BAINTON, D. F., and FARQUHAR, M. G. Origin of granules in polymorphonuclear leukocytes. *J. Cell Biol.,* 28:277 (1966).

BEAMS, H. W., and KESSEL, R. G. The Golgi apparatus: structure and function. *Int. Rev. Cytol.,* 23:209 (1968).

BERNHARD, W., and DE HARVEN, E. L'ultrastructure du centriole et d'autres éléments de l'appareil achromatique. *In* Proceedings of the Fourth International Conference on Electron Microscopy, Berlin, 1958. W. Bargmann, D. Peters, and C. Wolpers, editors. Berlin, Springer, vol. II (1960) p. 217.

CUNNINGHAM, W. P., MORRÉ, J. D., and MOLLENHAUER, H. H. Structure of isolated plant Golgi apparatus revealed by negative staining. *J. Cell Biol.,* 28:169 (1966).

DALTON, A. J., and FELIX, M. D. Cytological and cytochemical characteristics of the Golgi substance of epithelial cells of the epididymis—in situ, in homogenates, and after isolation. *Amer. J. Anat.,* 94:171 (1954).

FLICKINGER, C. J. The development of Golgi complexes and their dependence upon the nucleus in amebae. *J. Cell Biol.,* 43:250 (1969).

FRIEND, D. S., and FARQUHAR, M. G. Functions of coated vesicles during protein absorption in the rat vas deferens. *J. Cell Biol.,* 35:357 (1967).

FRIEND, D. S. and MURRAY, M. J. Osmium impregnation of the Golgi apparatus. *Amer. J. Anat.,* 117:135 (1965).

GALL, J. G. Centriole replication. A study of spermatogenesis in the snail *Viviparus. J. Biophysic. and Biochem. Cytol.,* 10:163 (1961).

MIZUKAMI, I., and GALL, J. Centriole replication. II. Sperm formation in the fern, *Marsilea,* and the cycad, *Zamia. J. Cell Biol.,* 29:97 (1966).

MURRAY, R. G., MURRAY, A. S., and PIZZO, A. The fine structure of mitosis in rat thymic lymphocytes. *J. Cell Biol.,* 26:601 (1965).

NEUTRA, M., and LEBLOND, C. P. The Golgi apparatus. *Sci. Amer.,* 220:100 (February, 1969).

ROBBINS, E., and GONATAS, N. K. The ultrastructure of a mammalian cell during the mitotic cycle. *J. Cell Biol.,* 21:429 (1964).

SZOLLOSI, D. The structure and function of centrioles and their satellites in the jellyfish *Phialidium gregarium. J. Cell Biol.,* 21:465 (1964).

WHALEY, W. G., DAUWALDER, M., and KEPHART, J. E. Golgi apparatus: influence on cell surfaces. *Science,* 175:596 (1972).

PLATE 4
The Interphase Nucleus and Nucleolus

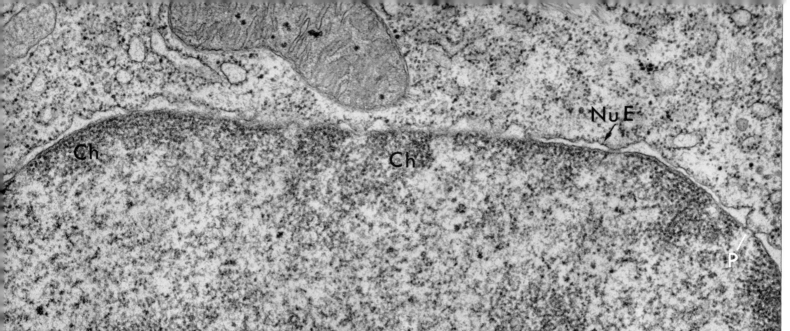

PLATE 4

The Interphase Nucleus and Nucleolus

The interphase nucleus gives little morphological evidence of the important activities occurring within it. It was once called the "resting" nucleus, but this name is highly unsuitable, because it is now known that during the periods between cell divisions the nucleus is exerting its greatest influence on events occurring in the cytoplasm. Furthermore, it is during the interphase period that the genetic material in the chromosomes is duplicated in preparation for the next cell division.

Electron microscope images have added surprisingly little to our knowledge of nuclear substructure apart from what they reveal regarding the nature of the nuclear envelope (NuE) and its pores (P), both discussed in Plate 2. Light microscope stains, combined with specific enzyme digestion, have localized deoxyribonucleic acid (DNA) in the chromatinic substance lying at the periphery of the nucleus as well as in the nucleolus, and they have further shown that the latter structure contains ribonucleic acid (RNA) as well. Careful comparison of light and electron micrographs established that DNA is localized in the central fibrillar zone (FZ) of the nucleolus, surrounded by a coarser granular zone (GZ) that is rich in RNA. The DNA-containing chromatin (the heterochromatin), or the bulk of it, appears as peripheral clumps of material (Ch) possessing a relatively uniform granular texture. Some also occurs near the nucleolus.

In cells such as the one shown here, in which synthetic activity in the cytoplasm is high, most of the nucleoplasm (the euchromatin) appears as fine granules and delicate filaments. These components are not detected with the light microscope because they are not sufficiently aggregated to be resolved or to combine with a noticeable amount of stain.

The functions of the interphase nucleus have been variously investigated. Certain large unicellular organisms, particularly amebae and the alga, *Acetabularia,* allow enucleation and nuclear transplantation experiments to be carried out. From such studies it has become clear that cells lacking a nucleus are unable to assimilate nutrients or to build and repair protoplasm. Furthermore, interspecific transfers of nuclei have shown that the nucleus soon exerts an influence on the cytoplasmic characteristics of the host cell.

This influence of the nucleus on the cytoplasm is demonstrably due to its control over the production of various kinds of RNA, which eventually reach the cytoplasm. As evidence of this, when cells are exposed to isotopically labeled precursors of RNA, the nuclei are labeled first and after only a short interval, whereas the cytoplasmic label appears much later. Moreover, destruction of RNA by the enzyme ribonuclease prevents regeneration of surgically excised parts of the cytoplasm.

The nuclear-derived RNA is not all of one type but consists rather of transfer, messenger, and ribosomal RNA's. Of these, ribosomal RNA is now known to be produced in the nucleolus. Treating cells with the antibiotic actinomycin D in low dosages has been found to interfere with the formation of DNA-dependent RNA, and when cells so treated are exposed to radioactive labeled precursors of RNA, no nucleolar labeling is observed. This result is in contrast to the marked nucleolar labeling in untreated cells. Dramatic confirmation of the role of the nucleolus in the production of RNA has been obtained also from the study of offspring from *Xenopus* (toad) parents each possessing only one of the normal pair of nucleoli. One quarter of the resulting embryos begin development without nucleoli. They live only a short time, but analysis of their RNA is possible, and, as expected, they lack ribosomal RNA.

This evidence correlates well with earlier observations that nucleoli undergo hypertrophy when cells are stimulated to greater protein synthesis, an event now equated with increased ribosomal activity. Thus when a nerve cell, as after nerve resection, is stimulated to regenerate a new axonal process and hence a large amount of cytoplasm, its nucleolus undergoes a 7- to 8-fold increase in volume (see text figure 4a).

In spite of this knowledge that the nucleolus produces ribosomal RNA, the RNA-rich granules of the nucleolus have not been shown to be equivalent to cytoplasmic ribosomes. As a matter of fact, the form in which various kinds of RNA are transferred to the cytoplasm is unknown.

The fibrillar zone of the nucleolus, where DNA

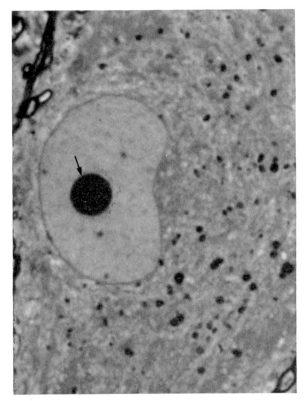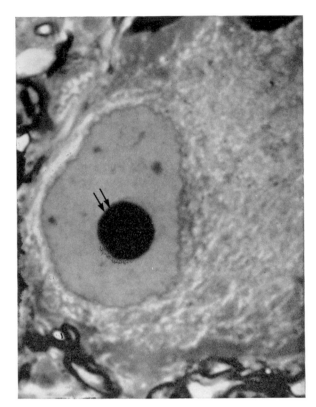

Text Figure 4a

Nuclei of neurons from the spinal cord of the frog are shown in two light micrographs at the same magnification. At left, a normal neuron displays the large dense nucleolus (arrow) characteristically present. At right, marked enlargement of the nucleolus (double arrow) is easily observed. The hypertrophy occurred after the surgical removal of the long axonal process of the cell. During the period of axonal regeneration that followed, the nucleolus was presumably stimulated to produce an extraordinary amount of ribosomal RNA, which would later have played a role in building up new protoplasmic proteins for the growing axon.

From the spinal cord of the frog
Magnification × 1,000

is located, represents part of one chromosome, a region known as the nucleolar organizer. In telophase, the last stage of cell division, the nucleus resumes synthesis of RNA and protein, i.e., processes interrupted at the beginning of the division cycle. Simultaneously some prenucleolar materials may be detected in the interchromosomal spaces. Subsequently these condense and associate with a specific chromosomal region, the nucleolar organizer. As a rule one organizer occurs in each haploid set of chromosomes, so that two nucleoli are formed in each somatic cell.

Concepts of chromosomal organization have depended primarily on studies of special cases in which interphase chromosomes retain at least a partially compact structure characteristic of mitotic chromosomes. The compact areas (e.g., in giant salivary gland chromosomes of dipterans; "lamp-brush" chromosomes of amphibian oocytes) stain positively for DNA and are con-sidered to be regions where the chromonemata, the filamentous threads of genetic material, have remained tightly coiled. Such areas are equated with heterochromatin and are regarded as relatively inactive regions of the genome. Less densely staining regions of the chromosomes (called euchromatin), which appear under the light microscope as "puffs" or "loops," contain some RNA, as evidenced by their staining properties. These are believed to be areas in which the chromonemata are uncoiled and where the genetic information is being transcribed, probably in the production of messenger RNA.

In the more common appearance of the interphase nucleus, such as that shown here, individual chromosomes lose their identity, and the peripheral clumps of material (Ch) associated with the interpore regions of the nuclear envelope probably correspond to the densely coiled regions of the salivary gland chromosome. Less dense

22

regions of the nucleoplasm, on the other hand, may be regarded as representing the uncoiled regions (euchromatin) where messenger RNA is being actively synthesized.

During cell division, the entire chromosome becomes a compact structure, and it is during that state that attempts have been made to determine how the nucleoprotein is arranged. Although the details are far from clear, it is evident that long threads of nucleoprotein are coiled in a complex way and that several orders of coiling

exist. Apparently the very small dense granules and strands in the denser chromatin (Ch) shown in this micrograph represent portions of the fundamental thread included in the section. Since interpretation of chromosomal structure from thin sections is far from simple, it is not surprising that the observed details have brought forth diverse interpretations.

From the pancreas of the frog
Magnification × 40,000

References

GALL, J. G., and CALLEN, H. G. H³ uridine incorporation in lampbrush chromosomes. *Proc. Nat. Acad. Sci. U.S.A., 48:*562 (1962).

GOLDSTEIN, L., and PLAUT, W. Direct evidence for nuclear synthesis of cytoplasmic ribose nucleic acid. *In* Cell Biology. L. Goldstein, editor. Dubuque, W. C. Brown (1966) p. 104.

———, and PRESCOTT, D. M. Proteins in nucleocytoplasmic interactions. *J. Cell Biol., 33:*637 (1967).

GRANBOULAN, N., and GRANBOULAN, P. Cytochimie ultrastructurale du nucléole. II. Etude des sites de synthèse du RNA dans le nucléole et le noyau. *Exp. Cell Res., 38:*604 (1965).

GURDON, J. B., and BROWN, D. D. Cytoplasmic regulation of RNA synthesis and nucleolus formation in developing embryos of *Xenopus laevis. J. Molec. Biol., 12:*27 (1965).

HÄMMERLING, J. Nucleo-cytoplasmic relationships in the development of *Acetabularia. Int. Rev. Cytol., 2:*475 (1953).

HARRIS, H. The reactivation of the red cell nucleus. *J. Cell Sci., 2:*23 (1967).

———, SIDEBOTTOM, E., GRACE, D. C., and BRAMWELL, M. E. The expression of genetic information: a study with hybrid animal cells. *J. Cell Sci., 4:*499, (1969).

JONES, K. W. The role of the nucleolus in the formation of ribosomes. *J. Ultrastruct. Res., 13:*257 (1965).

LAFONTAINE, J. G., and CHOUINARD, L. A. A correlated light and electron microscope study of the nucleolar material during mitosis in *Vicia faba. J. Cell Biol., 17:*167 (1963).

LITTAU, V. C., ALLFREY, V. G., FRENSTER, J. H., and MIRSKY, A. E. Active and inactive regions of nuclear chromatin as revealed by electron microscope autoradiography. *Proc. Nat. Acad. Sci. U.S.A., 52:*93 (1963).

MAGGIO, R., SIEKEVITZ P., and PALADE, G. E. Studies on isolated nuclei. II. Isolation and chemical characterization of nucleolar and nucleoplasm subfractions. *J. Cell Biol., 18:*293 (1963).

MARINOZZI, V. Cytochimie ultrastructurale du nucléole— RNA et protéines intranucléolaires. *J. Ultrastruct. Res., 10:*433 (1964).

McCONKEY, E. H., and HOPKINS, J. W. The relationship of the nucleolus to the synthesis of ribosomal RNA in Hela cells. *Proc. Nat. Acad. Sci. U.S.A., 51:*1197 (1964).

POLLISTER, A. W., SWIFT, H., and ALFERT, M. Studies on desoxypentose nucleic acid content of animal nuclei. *In* Cell Biology. L. Goldstein, editor. Dubuque, W. C. Brown (1968) p. 54.

PRESCOTT, D. M. RNA synthesis in the nucleus and RNA transfer to the cytoplasm in *Tetrahymena pyriformis. In* Biological Structure and Function. T. W. Goodwin and O. Lindberg, editors. New York, Academic Press, vol. II (1961) p. 527.

REICH, E., FRANKLIN, R. M., SHATKIN, A. J., and TATUM, E. L. Effect of actinomycin D on cellular nucleic acid synthesis and virus production. *Science, 134:*556 (1961).

RIS, H. Ultrastructure and molecular organization of genetic systems. *Canad. J. Genet. Cytol., 3:*95 (1961).

SIDEBOTTOM, E., and HARRIS, H. The role of the nucleolus in the transfer of RNA from nucleus to cytoplasm. *J. Cell Sci., 5:*351 (1969).

SWIFT, H. Nucleic acids and cell morphology in dipteran salivary glands. *In* Molecular Control of Cellular Activity. J. M. Allen, editor. New York, McGraw-Hill (1962) p. 73.

WALLACE, H., and BIRNSTIEL, M. L. Ribosomal cistrons and the nucleolar organizer. *Biochim. Biophys. Acta, 114:*296 (1966).

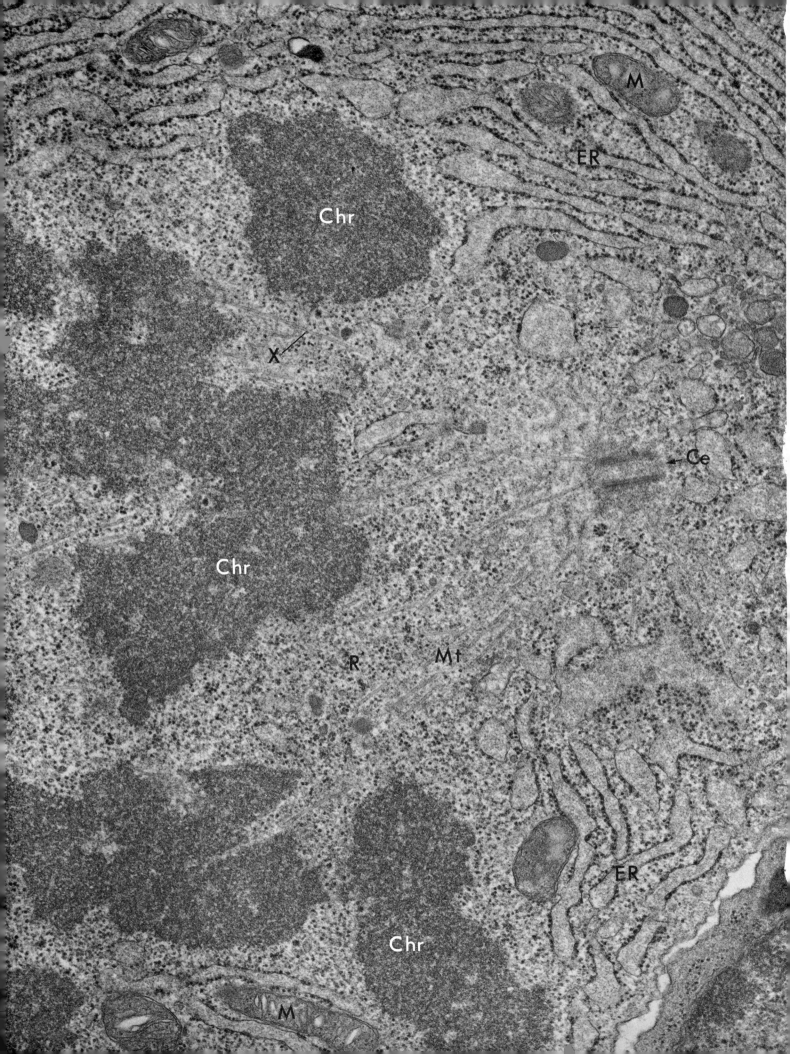

PLATE 5

The Mitotic Spindle

The spindle is a transient structure, yet one which is of prime importance in maintaining the continuity of genetic information from one cell generation to the next. It provides the mechanism that guides the chromosomes into ordered positions on the metaphase plate and normally effects the separation and movement of the daughter chromatids during anaphase. If spindle formation is prevented, as with the drug colchicine, or if the spindle is caused to disassemble, as with low temperatures, mitosis is blocked.

Normally the mitotic (or meiotic) movements occur only after complex preparatory changes have taken place in both the chromosomes and the centrioles. The latter changes are involved in the formation of the spindle apparatus.

The micrograph presented here has captured only a moment in cell division, and one during which chromosomes are united with the spindle in carrying out the characteristic mitotic movements best visualized by time-lapse cinematography. We can, however, see in this static picture evidence of earlier events as well as the structures involved in this dynamic process. The chromosomes (Chr) are now in a condensed state. Although one may assume that the essential genetic material is arranged in a compact and orderly way in these bodies, it remains extremely difficult from these images (as noted in Plate 4) to arrive at an exact description of their structure. Understanding the relationship among the minute components—described by some workers as 100 A filaments—presents a special challenge, because one is able to examine in thin sections only a small volume of the bulky chromosome.

A second important nuclear change is evident: the continuous envelope of the interphase nucleus (see also Plate 4) has disappeared, and the chromosomes are no longer "isolated" from the cytoplasm. Although remnants of the envelope have been followed through mitosis, none is clearly identifiable here. Only elements of the endoplasmic reticulum (ER), interspersed with mitochondria (M), lie at the perpihery of the cell. Near the end of mitosis, when the chromosomes of the daughter nuclei have aggregated at the poles of the spindles (telophase), a new envelope forms, apparently arising from cisternae of the adjacent ER. Thus the concept of the nuclear envelope as a specialized portion of the endoplasmic reticulum is given added credence.

While chromosomal movements are in progress, as at early anaphase shown here, the spindle plasm contains ribosome-like particles (R) intermingled with spindle microtubules (Mt). The particles are probably derived from the nucleoli, which disassembled in late prophase. The tubules, which are long, straight structures of uniform diameter, can be traced for a considerable distance in longitudinal section (as at X) and can also be identified as tiny rings in cross section (text figure 5a, arrowheads).

The existence of fibers within the spindle, as well as their function, is indicated by observations with the light microscope, especially in studies using polarized light. The form birefringence of the spindle points to the existence of asymmetric structural units oriented parallel to the spindle axis. When the changes in birefringence are observed during division, the distribution of these fibers in the spindle and the kinds of movements occurring can be analyzed.

Two kinds of fibers can be described, using light microscopy. The first appears to extend from one pole of the spindle to the other and is most properly referred to as a spindle fiber. A second kind reaches from the pole to the chromosome, to which it is attached at a special structure called a kinetochore; this is called a chromosomal fiber. The poles of the spindle are established during prophase when the centrioles in the cell center (cytocentrum) separate and migrate to antipodal positions with respect to the nucleus. It is from these poles that the spindle originates and becomes organized.

Although it is impossible to examine spindle fibers over their entire length in electron micrographs, short lengths of both types described above have been identified in thin sections, and the disposition of microtubules in them has been studied. Longitudinal sections of tubules have been found passing between the chromosomes at anaphase and have been located centrally between the chromosomal masses at telophase. In addition, spindle microtubules have been seen radiating from each polar centriole, as they do from the one (Ce) shown in this micro-

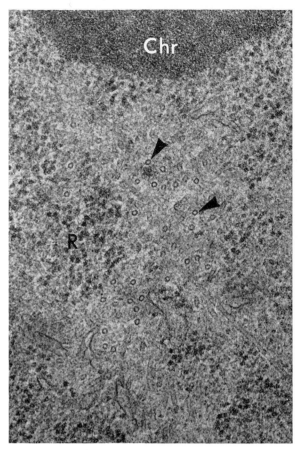

Text Figure 5a

When the mitotic spindle is cut in an equatorial section, cross sections of the spindle fibers (arrowheads) that run from the pole to the equator include circular profiles 240 A in diameter. The birefringent fibers of the spindle comprise then in part microtubules displaying a structural similarity to ciliary fibers (Plate 8) and neurotubules (text figure 45a). Part of a mitotic chromosome (Chr) and ribosome-like particles (R) may also be seen.

From the bone marrow of the rat
Magnification × 64,000

graph. It is probable that the microtubules do not extend without interruption from pole to pole within the spindle but rather consist of two sets that overlap in the fibers at the spindle equator. These are thought to be involved in spindle elongation, observed as an increase in the distance between the poles during anaphase and telophase. The attachment of microtubules to the chromosomes at the kinetochore has also been observed. These, along with spindle fibers, are believed to be instrumental in moving the chromosomes toward the poles. The mechanism involved in the earlier distribution of the chromosomes on the equatorial plate in metaphase is not in the least understood. Electron microscope studies

have therefore demonstrated that both spindle and chromosomal fibers comprise microtubules. It does not follow that the spindle fibers of classical light microscopy consist only of microtubules, for, in fact, they probably include adjacent spindle plasm as well. However, it seems appropriate to assign to the microtubules a central role in determining the form and functional characteristics of the spindle (see below).

Microtubules are so named because in cross section they appear as dense rings around less dense centers (text figure 5a, arrowheads). Similar structures (long, straight, slender rods, ca. 240 A in diameter) have now been described in a wide variety of cells and seem to be among the regularly occurring cell structures. Though they may often be found in orderly array, as in cilia (see Plate 8), they may assume a more random distribution. Their association with the cell center has been confirmed in a number of instances in which one end of a microtubule has been seen embedded in a centriole or a centriolar satellite. The spindle therefore appears to be a special arrangment of a cytoplasmic element that is widely, if not universally, present in cells. In this instance the microtubules assemble out of monomeric tubule protein within the nucleoplasm and cytoplasm to form the spindle, and after mitosis they apparently disassemble to become part of the cytoplasmic ground substance.

The occurrence of microtubules in motile cell organelles, such as cilia, and at sites where other cytoplasmic movements take place suggests that they play a significant role in protoplasmic movements in general. How they do this is the critical but unanswered question. One hypothesis currently in vogue is that they provide an elastic frame along which the motile force is exerted.

Isolation of large numbers of mitotic spindles from synchronously dividing populations of sea urchin eggs has provided sufficient material for biochemical analysis of spindle components. An important constituent is a protein that bears some similarity to actin of muscle (see Plate 38) and a close identity with a protein that has been isolated from cilia (see Plate 8) and identified with the microtubules. There is in the spindle, as in muscle and cilia, a protein possessing the ability to split enzymatically the high energy bond of adenosine triphosphate (ATP). This ATP-ase, located in some instances in small condensations in the tubule surface (see also Plate 8), would seem to be strategically placed either for the bending motion of cilia or for the gliding movements of cell

26

organelles relative to stationary microtubules. Although the process by which chemical energy is transformed into mechanical energy is not understood in any of these cases, the investigator may expect that the movement of the chromo- somes conforms to a pattern common to other types of cell movements.

From the submucosa of the rat intestine
Magnification $\times$ 40,000

References

BAJER, A. Notes on ultrastructure and some properties of transport within the living mitotic spindle. *J. Cell Biol., 33:*713 (1967).

BEHNKE, O. A preliminary report on "microtubules" in undifferentiated and differentiated vertebrate cells. *J. Ultrastruct. Res., 11:*139 (1964).

BORISY, G. G., OLMSTED, J. B., and KLUGMAN, R. A. *In vitro* aggregation of cytoplasmic microtubule subunits. *Proc. Nat. Acad. Sci. U.S.A., 69:*2890 (1972).

————, and TAYLOR, E. W. The mechanism of action of colchicine. Binding of colchicine-^{3}H to cellular protein. *J. Cell Biol., 34:*525 (1967).

————, and TAYLOR, E. W. The mechanism of action of colchicine. Colchicine binding to sea urchin eggs and the mitotic apparatus. *J. Cell Biol., 34:*535 (1967).

BOUCK, G. B., and BROWN, D. L. Microtubule biogenesis and cell shape in *Ochromonas.* I. The distribution of cytoplasmic and mitotic microtubules. *J. Cell Biol., 56:*340 (1973).

BROWN, D. L., and BOUCK, G. B. Microtubule biogenesis and cell shape in *Ochromonas.* II. The role of nucleating sites in shape development. *J. Cell Biol., 56:* 360 (1973).

BYERS, B., and ABRAMSON, D. H. Cytokinesis in HeLa: post-telophase delay and microtubule-associated motility. *Protoplasma, 66:*413 (1968).

HARRIS, P. Some observations concerning metakinesis in sea urchin eggs. *J. Cell Biol., 25* (No. 1, part 2): 73 (1965).

INOUÉ, S. Organization and function of the mitotic spindle. *In* Primitive Motile Systems in Cell Biology. R. D. Allen and N. Kamiya, editors. New York, Academic Press (1964) p. 549.

————, and BAJER, A. Birefringence in endosperm mitosis. *Chromosoma, 12:*48 (1961).

————, and SATO, H. Cell motility by labile association of molecules. The nature of mitotic spindle fibers and their role in chromosome movement. *J. Gen Physiol., 50* Suppl.:259 (1967).

JOHNSON, U. G., and PORTER, K. R. Fine structure of cell division in *Chlamydomonas reinhardi.* Basal bodies and microtubules. *J. Cell Biol., 38:*403 (1968).

MAZIA, D. How cells divide. *Sci. Amer., 205:*101 (September, 1961).

McINTOSH, J. R. The axostyle of *Saccinobaculus.* II. Motion of the microtubule bundle and a structural comparison of straight and bent axostyles. *J. Cell Biol., 56:*324 (1973).

————, and LANDIS, S. C. The distribution of spindle microtubules during mitosis in cultured human cells. *J. Cell Biol., 49:*468 (1971).

————, OGATA, E. S., and LANDIS, S. C. The axostyle of *Saccinobaculus.* I. Structure of the organism and its microtubule bundle. *J. Cell Biol., 56:*304 (1973).

MOOSEKER, M. S., and TILNEY, L. G. Isolation and reactivation of the axostyle. Evidence for a dynein-like ATPase in the axostyle. *J. Cell Biol., 56:*13 (1973).

PORTER, K. R. Cytoplasmic microtubules and their functions. *In* Principles of Biomolecular Organization. Ciba Foundation Symposium. G. E. W. Wolstenholme and M. O'Connor, editors. London, J. and A. Churchill, Ltd. (1966) p. 308.

————, and MACHADO, R. D. Studies on the endoplasmic reticulum. IV. Its form and distribution during mitosis in cells of onion root tip. *J. Biophysic. and Biochem. Cytol., 7:*167 (1960).

ROTH, L. E., and DANIELS, E. W. Electron microscopic studies of mitosis in amebae. *J. Cell Biol., 12:*57 (1962).

SAKAI, H. Studies on sulfhydryl groups during cell division of sea urchin eggs. VIII. Some properties of mitotic apparatus proteins. *Biochim. Biophys. Acta, 112:*132 (1966).

SHELANSKI, M. L., and TAYLOR, E. W. Isolation of a protein subunit from microtubules. *J. Cell Biol., 34:*549 (1967).

SLAUTTERBACK, D. B. Cytoplasmic microtubules. I. Hydra. *J. Cell Biol., 18:*367 (1963).

STEPHENS, R. E. On the structural protein of flagellar outer fibers. *J. Molec. Biol., 32:*277 (1968).

————. Factors influencing the polymerization of outer fiber microtubule protein. *Quart. Rev. Biophys., 1:*377 (1969).

————. Microtubules. *In* Biological Macromolecules. V. S. M. Timasheff and G. D. Fasman, editors. New York, Marcel Dekker (1971) p. 355.

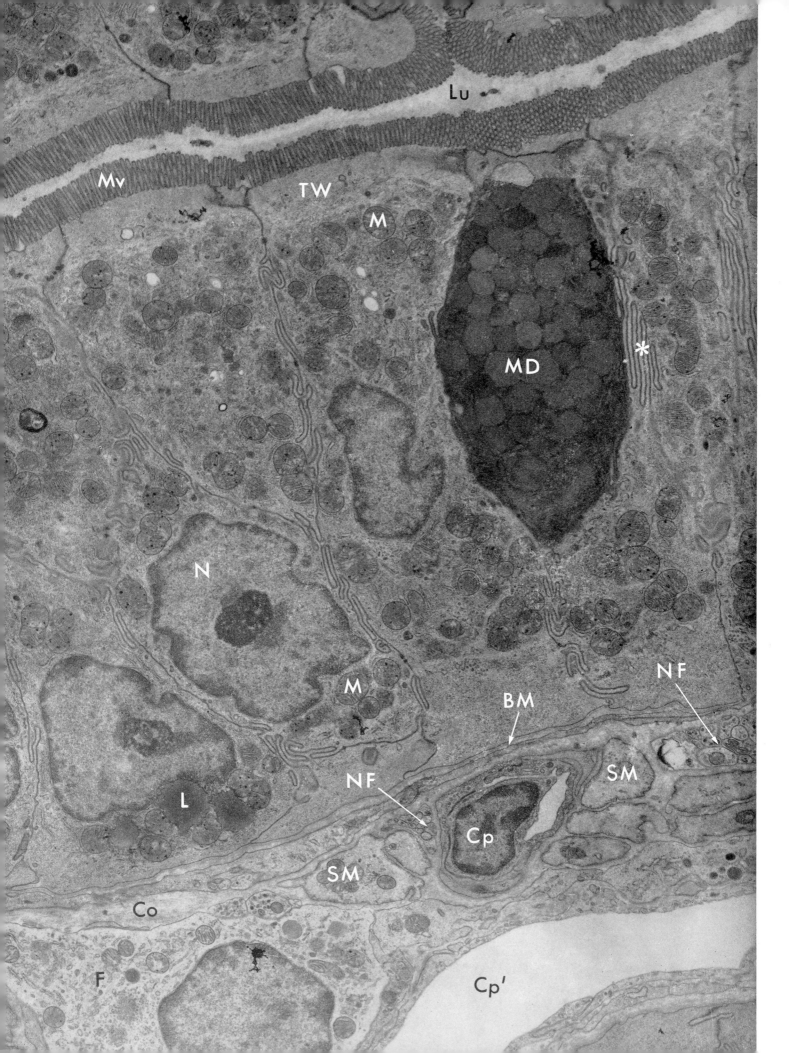

PLATE 6
The Intestinal Epithelium

In multicellular animals the association of cell units into tissues is the rule. Epithelia, in general, form the covering and lining tissues of the animal body. One type, a simple columnar epithelium, can be examined in this plate, and several of its characteristics noted. The individual cells, which are here roughly cylindrical units, are so closely packed together in hexagonal array that little intercellular space occurs between the plasma membranes of adjacent lateral cell surfaces. Instead the epithelial cells at their lateral surfaces are attached to each other by means of certain specializations of the cortex and the plasma membrane. These take the form of interdigitating folds of adjacent cell surfaces (*) or appear as specialized adhesion areas, called tight junctions, terminal bars, and desmosomes (see Plates 7 and 9). Together these structures form the junctional complexes, which seal off the intestinal lumen from the intercellular spaces within and beneath the epithelium.

Two cell types, each performing its special function, have differentiated within this epithelium, the predominant one being the columnar absorptive cell. This tall cell extends from the basement membrane (BM)[1] to the lumen of the gut (Lu). It has long been recognized that the apical surfaces of the absorptive cells have a specialized structure, called the striated border, but the nature of this specialization was not clear until the tissue was examined under the electron microscope. Then this striated border of the light micrographs was resolved into a series of finger-like cytoplasmic projections or microvilli (Mv), each limited by the plasma membrane of the absorptive cell (see Plate 7). The material within the microvilli is continuous with that of a fibrillar ectoplasmic zone, the so-called terminal web (TW), from which the common cytoplasmic organelles are excluded. The cytoplasm subjacent to the terminal web is, in contrast, crowded with the usual cytoplasmic constituents, such as mitochondria (M). The region below the nucleus (N),

though less extensive, is also populated by mitochondria (M) and by lipid droplets (L).

The intestinal epithelium also includes mucus-secreting goblet cells. The "goblet" portion of one of these, which appears in this micrograph, is crowded with numerous mucous droplets (MD). The formation and discharge of mucus, a complex polysaccharide that coats and protects the surface of the epithelium, is considered in Plate 12.

Nutrient materials from the lumen of the gut do not pass between the cells of the epithelium, but rather through their apical surfaces and are transported thence to their basal or lateral surfaces. Both the physical process of diffusion and chemical reactions governed by enzymes (active transport) are important in moving nutrients from the lumen to the underlying tissues.

The basement membrane (also called the basal lamina) forms a barrier between the epithelium and the supporting tissue beneath it. Such an arrangement, that is, the presence of a thin amorphous polysaccharide-rich layer between different types of tissues, is almost universal. It seems likely that the basement membrane is formed by or at least with the participation of the epithelium itself. A more complete discussion of this structure is provided in the legend of Plate 28.

The lamina propria, the supporting layer underlying the intestinal epithelium, is a pliable yet strong layer of connective tissue within which are embedded blood and lymph vessels, nerves, and muscles. Connective tissues, in contrast to epithelia, are characterized by the presence of substantial amounts of various intercellular substances. The latter consist primarily of collagen fibrils (Co), forming a flexible and tough supporting element, together with an amorphous ground substance rich in mucopolysaccharide. The ground substance holds the fibrous elements in place and is the medium of the intercellular spaces through which metabolites diffuse. Fibrils and ground substance are produced by fibroblasts (F). Basement membranes seem, on the other hand, to be produced by the cells they surround or underlie.

The lamina propria is heavily vascularized. Blood capillaries, both in cross section (Cp) and

---

[1] The term *basement membrane,* which was used in the second edition of this book, is now being replaced in histological literature by *basal lamina*. These two terms will be used interchangeably in this edition, since both occur in the literature cited.

29

in longitudinal section (Cp′), are evident. They are lined by a thin epithelium, called an endothelium, which is underlain in turn by a basement membrane. Details of capillary structure and examples of their structural diversity may be found in a number of tissues illustrated in this collection, especially Plates 10, 24, 25, 26, 36, and 39. A second type of vessel, quite similar to a blood capillary, occurs in the lamina propria. This is the lacteal, a lymph capillary, which carries away absorbed fat in the form of small droplets, the chylomicrons.

The intestinal epithelium and lamina propria are molded into cylindrical or club-shaped projections, the villi, which impart a velvety appearance to the luminal lining when viewed with the low power of a hand lens. These villi are able to move slowly, due to the presence within them of smooth muscle cells (SM). Such involuntary movements aid in exposing the intestinal surface to hydrolyzed food substances and in moving circulating fluids away from the intestine.

Minute autonomic nerve fibers (NF) are frequently seen coursing through the lamina propria. These are extremely slender processes (~ 0.2–0.5μ in diameter) of nerve cells located elsewhere in aggregations called ganglia.

This micrograph illustrates the nature of cell associations in epithelial and connective tissues. Both types of tissue, and especially the connective tissue, are a composite of various kinds of cells, each specialized for particular function, yet both tissues as such possess easily recognizable properties that permit their classification and identification.

From the duodenum of the bat (*Myotis lucifigus*)
Magnification $\times$ 7,500

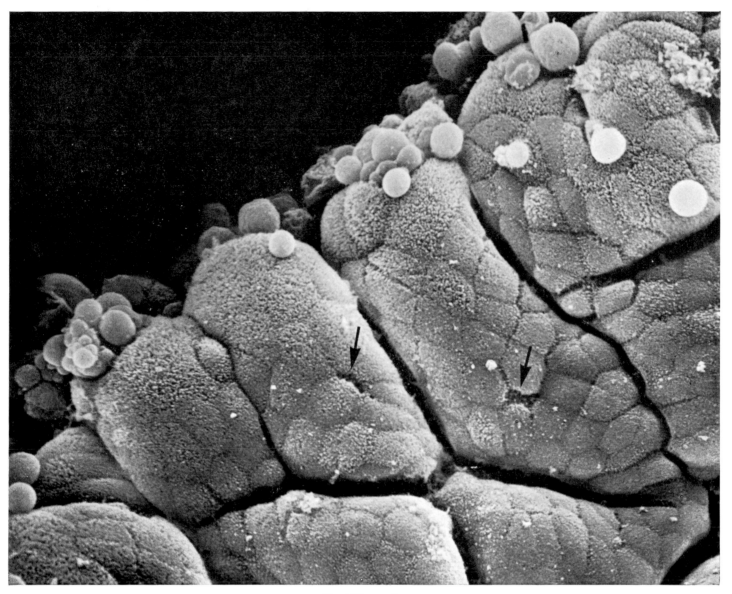

Text Figure 6a

The villi that line the inner surface of the small intestine in large numbers are small, tongue-shaped structures about 0.5 to 1.0 mm in diameter at their bases and 1.0 to 1.5 mm in height. In life they project vertically from the wall of the gut into the gut's lumen, but are able to bend slightly from the vertical position as well as to lengthen and shorten. Their presence obviously increases the inner surface area of the gut and thus facilitates absorption.

This scanning electron micrograph depicts a portion of the flat surface of one villus; its upper edge runs diagonally across the upper left. The observer is looking directly at the free surfaces of the cells making up the columnar epithelium and the cells' dense cover of microvilli (Plates 6 and 7). The individual cells, mostly with five or six sides, can be identified easily. The total surface seems to be divided into zones by fissures or sulci. The continuity of the epithelium is not interrupted at these fissures but extends around their surfaces. One assumes that

these "negative folds" in the surface provide for the small extensions and bendings of the villi in life.

As is now well known, the epithelium (the mucosa) of the intestine and its villi is constantly renewed. Cells that form in the crypts of Lieberkühn flow slowly toward the apical ridges of the villi, from which they are ultimately shed into the cavity of the gut. This latter event is represented here by the spheroidal (dying) cells and cell fragments arrayed along the free edge of the villus.

At a few points in the free surface of the mucosa one can recognize pits or small depressions where the surface texture changes (arrows). These mark the locations and free surfaces of goblet cells, which, unlike the adsorptive cells, possess only a few relatively short microvilli (see text figure 6b).

From the intestine of the rat
Magnification × 1,000

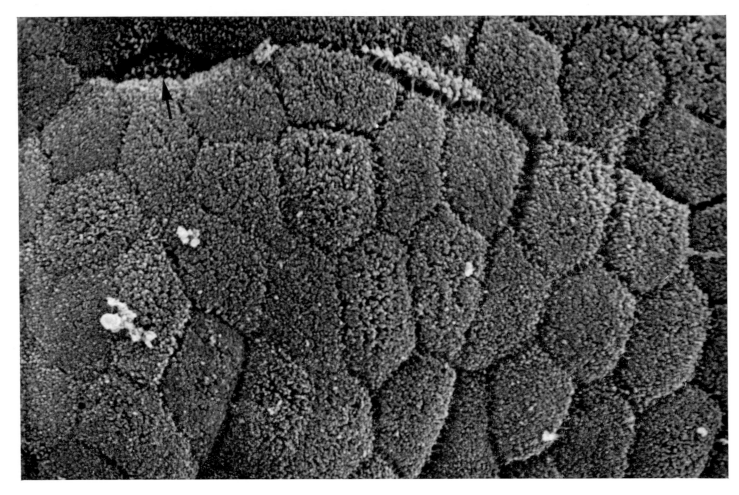

Text Figure 6b

In a scanning microscope picture taken at a higher magnification than that of text figure 6a, it is possible to see the individual microvilli that cover the apical surfaces of the intestinal epithelial cells. From such images as this, one perhaps gains a better impression of the huge numbers of microvilli involved than one does from sections (e.g., Plate 7) and also one can see how closely packed the microvilli are in covering the cell surfaces. Only the tips of these microvilli show. The outlines of the individual cells are easily identified. Only occasionally is their pavement-like pattern interrupted to permit a goblet cell to gain access to the surface (at arrow). The surface of the goblet cell, with its relatively few short microvilli, contrasts strikingly with that of the adsorptive cells.

From the intestine of the rat
Magnification × 2,000

ANDERSON, J. H., and WITHERS, R. H. Scanning electron microcsope studies of irradiated rat intestinal mucosa. *In* Scanning Electron Microscopy/1973. O. Johari and I. Corvin, editors. Chicago, IIT Research Institute (1973) p. 565.

DEANE, H. W. Some electron microscopic observations on the lamina propria of the gut, with comments on the close association of macrophages, plasma cells, and eosinophils. *Anat. Rec., 149:*453 (1964).

GREY, R. D. Morphogenesis of intestinal villi. I. Scanning electron microscopy of the duodenal epithelium of the developing chick embryo. *J. Morphol., 137:*193 (1972).

LEBLOND, C. P., and MESSIER, B. Renewal of chief cells and goblet cells in the small intestine as shown by radioautography after injection of thymidine-H³ into mice. *Anat. Rec., 132:*247 (1958).

———, and STEVENS, C. E. The constant renewal of the intestinal epithelia in the albino rat. *Anat. Rec., 100:* 357 (1948).

———, and WALKER, B. E. Renewal of cell population. *Physiol. Rev., 36:*255 (1956).

NEUTRA, M., and LEBLOND, C. P. Radioautographic comparison of the uptake of galactose-H³ and glucose-H³ in the Golgi region of various cells secreting glycoproteins or mucopolysaccharides. *J. Cell Biol., 30:*137 (1966).

PALAY, S. L., and KARLIN, L. J. An electron microscopic study of the intestinal villus. I. The fasting animal. *J. Biophysic. and Biochem. Cytol., 5:*363 (1959).

RAO, N. S., and WILLIAMS, W. A. Normal and ischaemic jejunal mucosa of mice. Scanning electron microscope study. *J. Microsc. (Paris), 15:*219 (1972).

TONER, P. G., and CARR, K. E. The use of SEM in the study of the intestinal villi. *J. Pathol., 79:*611 (1969).

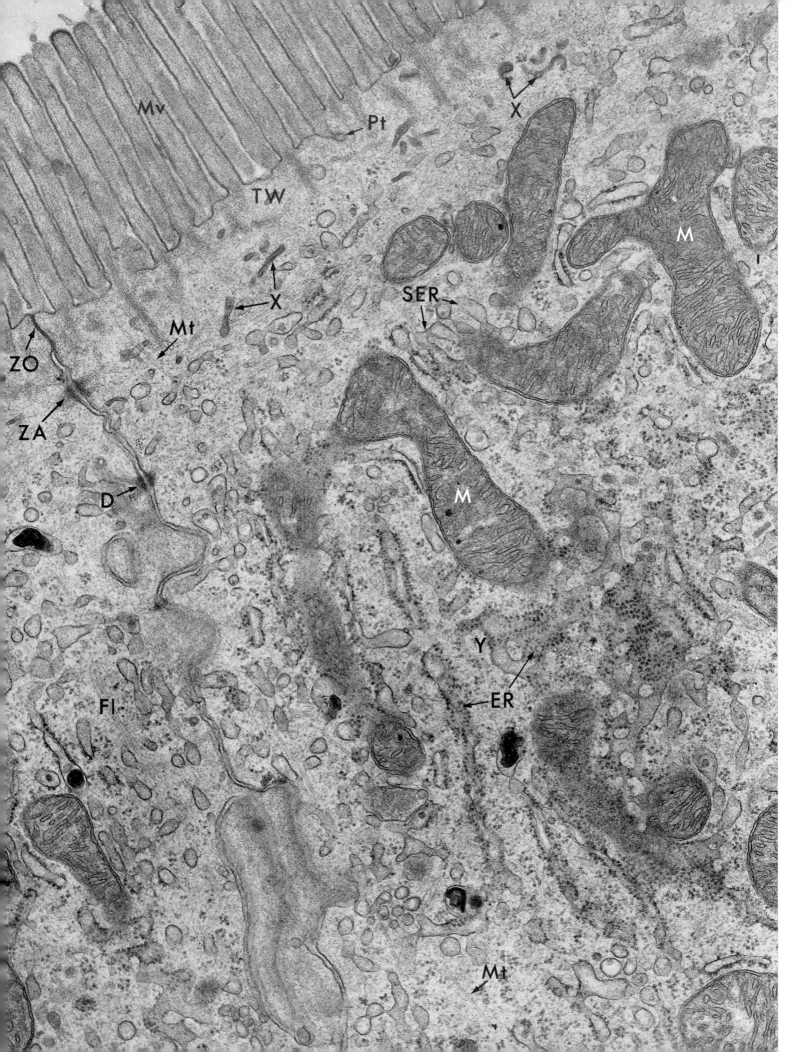

PLATE 7

Columnar Absorptive Cells

The apical surface of columnar absorptive cells is an important area of contact between the organism and its environment. It is here that food materials, after undergoing extracellular digestion (hydrolysis), really enter the organism.

The absorbing surface is enlarged by folds, both macroscopic and microscopic, and, as already shown in Plate 6, studies of this tissue by electron microscopy have revealed the presence here of numerous fingerlike projections, the microvilli (Mv). Fine filaments constitute the cores of these projections and extend into the terminal web (TW) area. A close examination of the microvillar profiles demonstrates the existence of a filamentous enteric surface coat external to, but intimately associated with, the plasma membrane (text figure 7a). This layer, which consists of mucopolysaccharide, is believed to be produced by the absorptive cell itself and to be an integral part of the cell surface. The coat probably functions as a trap that can concentrate ions and other charged particles to be absorbed by the cells (see also Plate 13).

Far down between the microvilli the plasma membrane can form pits (Pt) that penetrate the terminal web. Irregular profiles of pits or apical vesicles (X) derived from them are easily identified within the web. Dense material of unknown nature is frequently found within these vesicles.

The lateral surfaces of the absorptive cells are closely united by components of the junctional complex. The tight junction (or *zonula occludens*, ZO), involving fusion between adjacent plasma membranes, is located closest to the free surface of the epithelium. Its structure is seen more clearly in text figure 7a. Beneath it may be seen the terminal bar (or *zonula adhaerens*, ZA). Each of these constitutes a continuous ring encircling each cell of the epithelium at this level. In profile the terminal bar appears somewhat similar to the desmosome (D) or *macula adhaerens*. The latter, however, is only a small disk-shaped area where adjacent plasma membranes adhere. It is described in detail in the legend of Plate 9. Where they are not joined by specialized structures, the plasma membranes of adjacent cells are separated by a zone of constant width and low density. Although no structure is discerned within this gap, it is demonstrably a layer of material rich in polysaccharides.

The cytoplasm of absorptive cells subjacent to the terminal web is the site of great activity in the uptake and transport of metabolites. The fine filaments (Fl) and microtubules (Mt) present in the ground substance of the cytoplasm are probably visible manifestations of a cytoskeletal organization influential in maintaining the elongated shape of the cells and in facilitating diffusion. Mitochondria (M) packed with cristae are numerous. Their prominence is to be expected, since energy is required for transporting foodstuffs across the epithelium.

Also conspicuous in this region is a well-developed endoplasmic reticulum. Here the ER is quite different in form from that seen in the plasma and pancreatic cells (Plates 1, 2, and 11). Some portions of this system are studded with ribosomes (ER), but the membranes, rather than appearing as flattened cisternae, display a reticulate structure of anastomosing tubules. The ER membranes are also found free of ribosomes (SER), and in such regions are more vesicular in nature. It is usually a simple matter, as in this micrograph, to find places where rough- and smooth-surfaced membranes are continuous (Y). This is taken as evidence that the two forms are part of the same intracellular membrane system and suggests that one may transform into the other.

In studying this micrograph we see then a number of specialized structures associated with the cell surface and the apical cytoplasm. The extent and manner in which each may play a role in absorption and transport of molecules now become subjects of some interest. In the investigation of these phenomena it is rewarding to study the movement of emulsified fat across the columnar epithelial cells, because fat droplets are preserved and can be readily identified in electron micrographs when osmium tetroxide is used as a fixative.

In such investigations it has been observed that when emulsified corn oil is introduced into the lumen of the rat digestive tract, small droplets of fat appear soon afterward, mainly within the cavities of the smooth-surfaced endoplasmic reticulum (text figure 7b). With time the droplets

35

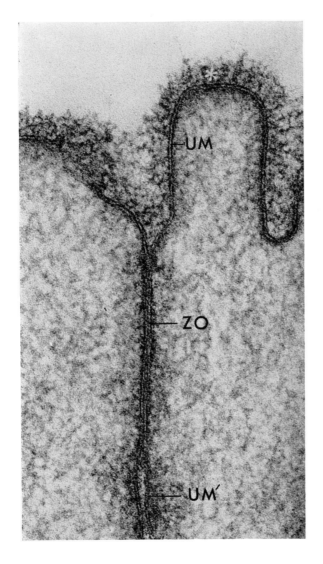

Text Figure 7a

The surfaces of absorptive cells are limited by a trilaminar unit membrane both at their free (UM) and on their lateral (UM′) surfaces. An enteric surface coat (*) of polysaccharide material covers the folds that project into the lumen of the stomach. At the lateral cell surfaces, the outer dense lamellae of the plasma membranes of adjacent cells are united to form a tight junction or *zonula occludens* (ZO). This structure is believed to act as a seal, preventing the diffusion of even small molecules through the epithelium by an intercellular route.

From the stomach of the bat
Magnification × 180,000

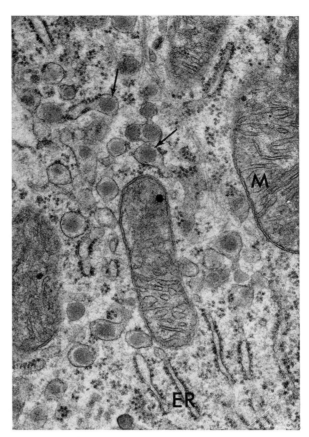

Text Figure 7b

The supranuclear cytoplasm of the columnar absorptive cell depicted here was fixed following injection of emulsified corn oil into the lumen of the intestine. Vesicles of the endoplasmic reticulum contain fat droplets (arrows) resynthesized from monoglycerides and fatty acids absorbed from the gut lumen. Tubules of rough-surfaced endoplasmic reticulum (ER) give rise to the vesicles, which are largely free of ribosomes. Mitochondria (M) are numerous.

From the intestine of the rat
Magnification × 36,000

become larger and increase in number as more and more of the rough ER is transformed into the smooth form. There is now convincing evidence that the membranes of the SER are the sites at which triglycerides (fats) are synthesized from monoglycerides and fatty acids that have diffused into the cell from the gut lumen. The droplets coated with protein and surrounded by a membrane seem then to move toward the basilateral surfaces of the columnar cells, where they are ejected, free of membrane, into the intercellular spaces. Their elimination from the cell is believed to involve momentary union of their surrounding membrane with the plasma membrane and the opening of the vesicle so that its inner cavity is for a brief time continuous with the intercellular space.

The pits (Pt) and the apical vesicles (X) derived from them constitute a system that apparently functions in the uptake and destruction of miscellaneous particulate materials that find their way into the deep recesses between the microvilli. When electron dense markers are added to corn oil in the form of fine colloidal suspensions, they are detected in pits and apical vesicles but never reach the internal phase of the SER. In time the apical vesicles are transformed into lysosomes, within which the intruding material is hydrolyzed or otherwise rendered harmless. A similar mechanism for dealing with large foreign molecules has been observed in the absorptive cells of the kidney (see Plate 24).

While diffusion processes adequately account for movement across the brush border (microvilli and terminal web) of digested fats in the form of monoglycerides, absorption of proteins and sugars requires energy and is therefore said to involve "active" transport processes. Specific knowledge of the molecular architecture, that is, the complex lipoprotein plasma membrane of the cell surface, is required to understand such processes (text figure 7a). It is thought that carrier molecules exist and that their properties account for the selectivity shown by the membrane in its ability to distinguish between quite similar molecular species.

From the jejunum of the rat
Magnification $\times$ 42,000

References

BENNETT, G. Migration of glycoprotein from Golgi apparatus to cell coat in the columnar cells of the duodenal epithelium. *J. Cell Biol.*, *45*:668 (1970).

———, and LEBLOND, C. P. Formation of cell coat material for the whole surface of columnar cells in the rat small intestine, as visualized by radioautography with L-fucose-³H. *J. Cell Biol.*, *46*:409 (1970).

BONNEVILLE, M. A., and WEINSTOCK, M. Brush border development in the intestinal absorptive cells of *Xenopus* during metamorphosis. *J. Cell Biol.*, *44*:151 (1970).

CARDELL, R. R., JR., BADENHAUSEN, S., and PORTER, K. R. Intestinal triglyceride absorption in the rat. *J. Cell Biol.*, *34*:123 (1967).

CRANE, R. K. Structural and functional organization of an epithelial brush border. *In* Symposia of the International Society for Cell Biology. Volume 5, Intercellular Transport. K. B. Warren, editor. New York, Academic Press (1966) p. 71.

FARQUHAR, M. G., and PALADE, G. E. Junctional complexes in various epithelia. *J. Cell Biol.*, *17*:375 (1963).

FRIEND, D., and GILULA, N. B. Variations in tight and gap junctions in mammalian tissues. *J. Cell Biol.*, *53*: 758 (1972).

ISSELBACHER, K. J. Biochemical aspects of fat absorption. *Gastroenterology, 50*:78 (1966).

ITO, S. The enteric surface coat on cat intestinal microvilli. *J. Cell Biol.*, *27*:475 (1965).

KAYE, G. I., WHEELER, H. O., WHITLOCK, R. T., and LANE, N. Fluid transport in the rabbit gallbladder. A combined physiological and electron microscopic study. *J. Cell Biol.*, *30*:237 (1966).

LENTZ, T. L., and TRINKAUS, J. P. Differentiation of the junctional complex of surface cells in the developing *Fundulus* blastoderm. *J. Cell Biol.*, *48*:455 (1971).

PALAY, S. L., and KARLIN, L. J. An electron microscopic study of the intestinal villus. II. The pathway of fat absorption. *J. Biophysic. and Biochem. Cytol.*, *5*:373 (1959).

QUINTON, P. M., and PHILPOTT, C. W. A role for anionic sites in epithelial architecture. Effects of cationic polymers on cell membrane structure. *J. Cell Biol.*, *56*:787 (1973).

REVEL, J.-P., and ITO, S. The surface components of cells. *In* The Specificity of Cell Surfaces. B. Davis and L. Warren, editors. Englewood Cliffs, New Jersey, Prentice-Hall (1967) p. 211.

SENIOR, J. R. Intestinal absorption of fats. *J. Lipid Res.*, *4*:495 (1964).

STAEHELIN, L. A. Further observations on the fine structure of freeze-cleaved tight junctions. *J. Cell Sci.*, *13*:(in press) (1973).

———, MUKHERJEE, T. M., and WILLIAMS, A. W. Freeze-etch appearance of the tight junctions in the epithelium of small and large intestine of mice. *Protoplasma, 67*:165 (1969).

STIRLING, C. E. High-resolution radioautography of phlorizin-³H in rings of hamster intestine. *J. Cell Biol.*, *35*:605 (1967).

———, and KINTER, W. B. High-resolution radioautography of galactose-³H accumulation in rings of hamster intestine. *J. Cell Biol.*, *35*:585 (1967).

STRAUSS, E. W. Electron microscopic study of intestinal fat absorption in vitro from mixed micelles containing linolenic acid, monoolein, and bile salt. *J. Lipid Res.*, *7*:307 (1966).

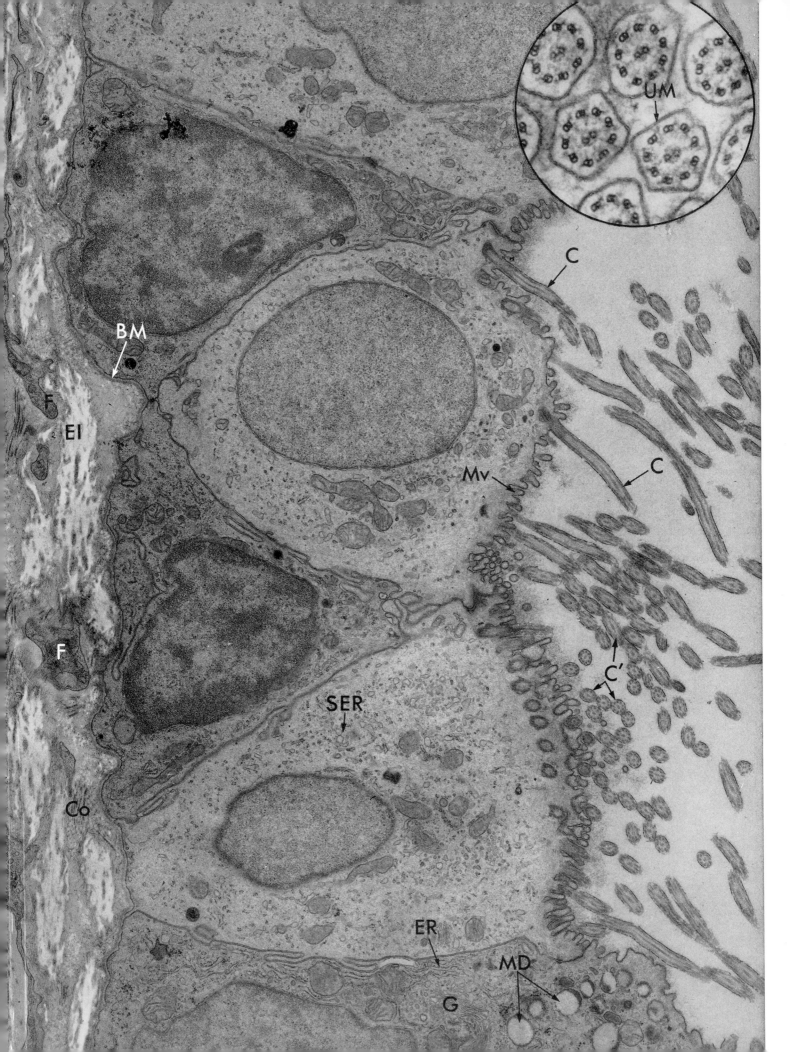

PLATE 8

The Ciliated Epithelium of the Trachea

The removal of foreign particles from the trachea is accomplished by the lashing motion of cilia, which are long, motile extensions from the free surfaces of certain epithelial cells. In this micrograph a few cilia (C) are seen in longitudinal section extending from the apical pole of epithelial cells. Because of their length, slenderness (250 nm in diameter), and contorted form, most of the cilia are included only in part in the section and so appear variously in oblique and cross section (C'). Like the short microvilli (Mv), which also project from the tracheal cells, cilia are covered by the three-layered unit membrane (UM, inset). Unlike the microvilli, however, cilia have a complex inner organization, which is shared in common by all cilia and flagella, whether from plant or animal cells. The nature of this organization is shown best in cross sections (inset). Thus one sees the "figure-8" profiles of nine double filaments or microtubules in a peripheral ring enclosing a central pair. This entire array of microtubules constitutes what has come to be known as the 9 + 2 complex or axoneme. The filamentous nature of these structural units is deduced from longitudinal sections. At the base of each cilium is a basal body, which closely resembles a centriole. Since the 9 + 2 complex was first discovered, more detailed descriptions of the various ciliary components have been published, and several mechanisms of ciliary motion have been proposed, based on the structural features revealed by the electron microscope. At this time the mechanism of movement and the manner in which the movements are coordinated into wave forms have not been fully explained.

Progress has been made, however, in isolating cilia in quantities that allow physical and chemical analysis of their components. By "chemical dissection" it has been possible to isolate the axoneme free of the limiting plasma membrane and relatively free of its surrounding matrix. Further extraction of the axoneme has led to isolation of a protein, dynein, which possesses ATPase activity. This fraction and its activity are of special interest because it is believed that splitting of the high energy phosphate bond of ATP provides energy for the movement of cilia as it does for the contraction of muscle (see Plates 38–40).

Electron microscopic examination of the axonemes following extraction of dynein reveals that the "arms" of the outer nine fibrils have disappeared. Normally the "arms," extending like prongs from the side of each doublet toward the one next, are responsible for the assymetry of the nine doublets. (The arms may be detected on close inspection of the inset.) If subsequently the dynein is recombined with the extracted axonemes, the "arms" are restored to their normal positions. It appears then that energy for ciliary movement is released near the nine double tubules, which probably serve simply as an elastic frame for the cylindrical shape of the cilium. How this energy acts to distort or bend this frame has yet to be discovered.

Further study of the nine outer doublets has shown that they are made up of a protein that resembles in a number of ways the protein actin, known to play an important role in the contraction of muscle. Although it is certain that the mechanisms of movement in muscle and cilia are not homologous, it seems likely that fundamental similarities exist.

In the tracheal epithelium the ciliated cells alternate with nonciliated, mucus-secreting cells, in the apical zones of which secretion droplets (MD) may be observed. The cytoplasm of the mucous cells is closely packed with organelles—mitochondria, endoplasmic reticulum (ER)—whereas in that of the ciliated cells, from which no product is exported, components other than mitochondria are rather sparsely represented. For example, the endoplasmic reticulum of the latter cells consists in the main of small smooth-walled vesicles (SER), and the Golgi complex (next to letters ER) is not as extensive as that which characterizes the secretory cells (G).

Although both ciliated and mucous cells extend the full height of the epithelium, slightly oblique sections may obscure this fact. In this micrograph some of the cells appear to be arranged in layers, while others are clearly part of a simple epithelium. The disposition of the nuclei at two or more levels within epithelia in which all components are in contact with the basement membrane accounts for use of the adjective "pseudostratified" to describe this epithelium.

Beneath this epithelium, as is true for epithelia in general, there is a thin basement membrane (BM), now more commonly called the basal lamina, which is supported in turn by rather substantial layers of fibrous connective tissue. Portions of fibrocytes (F) lie among collagenous (Co) and elastic (El) fiber bundles abundant in the connective tissue ground substance. It is this layer of connective tissue that is the major component of the thick basement lamella evident in the light microscope image of tracheal tissue.

From the trachea of the bat
Magnification × 13,000
Inset from the trachea of the rat
Magnification × 97,000

Text Figure 8a

Scanning Image of Tracheal Epithelium

Scanning Image of Tracheal Epithelium
(Text Figure 8a)

This scanning electron micrograph provides a surface view of the tracheal lining. The epithelium is comprised of two types of cells, already identified in Plate 8. The predominant type has numerous cilia, which give a shag-rug texture to the lining. The flexible cilia have been fixed while in various positions. Their inner structure is not revealed, and their certain identification depends upon sectioned specimens (Plate 8, inset) examined by transmission microscopy or upon observations of living cells. Nonetheless, the scanning image demonstrates, directly and clearly, the surface features of the epithelium, short-circuiting the necessity for reconstruction from serial sections. For example, one sees immediately the high density of the cilia and the distribution of ciliated versus nonciliated cells.

The nonciliated units, buried among the thick carpet of cilia, are small groups of mucous cells, the dome-shaped free surfaces of which are covered by short projections. In certain instances, the surfaces of these cells are smooth or wrinkled, and the microvillus-type projections are not seen. This latter image may represent an active secretory stage in which mucus is being emitted. The free cell surface may modulate between these two forms, secreting in a cyclical fashion. Alternatively, the secretory act may be a unique event for the mature cell, after which it dies, and certain of its fellows, arising from cell divisions, go on to maturity. Information on the dynamics of the epithelial population in health and disease comes within easy reach of the investigator using the scanning electron microscope.

From the trachea of the rat
Magnification $\times$ 4,200

References

BARBER, V. C., and BOYDE, A. Scanning electron microscopic studies of cilia. *Z. Zellforsch. Mikrosk. Anat., 84:*269 (1968).

FAWCETT, D. W. Cilia and flagella. *In* The Cell. J. Brachet and A. E. Mirsky, editors. New York, Academic Press, vol. II (1961) p. 217.

————, and PORTER, K. R. A study of the fine structure of ciliated epithelia. *J. Morph., 94:*221 (1954).

GIBBONS, B. H., and GIBBONS, I. R. Flagellar movement and adenosine triphosphatase activity in sea urchin sperm extracted with Triton-X 100. *J. Cell Biol., 54:*75 (1972).

GIBBONS, I. R. The relationship between the fine structure and direction of beat in gill cilia of a lamellibranch mollusc. *J. Biophysic. and Biochem. Cytol., 11:*179 (1961).

————. Studies on the protein components of cilia from *Tetrahymena pyriformis. Proc. Nat. Acad. Sci. USA, 50:*1002 (1963).

————, and ROWE, A. J. Dynein: a protein with adenosine triphosphatase activity from cilia. *Science, 149:* 424 (1965).

GREENWOOD, M. F., and HOLLAND, P. The mammalian respiratory tract surface. A SEM study. *Lab. Invest., 27:*296 (1972).

GRIMSTONE, A. V., and KLUG, A. Observations on the substructure of flagellar fibres. *J. Cell Sci., 1:*351 (1966).

RENAUD, F. L., ROWE, A. J., and GIBBONS, I. R. Some properties of the protein forming the outer fibers of cilia. *J. Cell Biol., 36:*79 (1968).

SATIR, P. Studies on cilia. II. Examination of the distal region of the ciliary shaft and the role of the filaments in motility. *J. Cell Biol., 26:*805 (1965).

————. Studies on cilia. III. Further studies on the cilium tip and a "sliding filament" model of ciliary motility. *J. Cell Biol., 39:*77 (1968).

STEPHENS, R. E. Studies on the development of the sea-urchin (*Strongylocentrotus droebachiensis.* III. Embryonic synthesis of ciliary proteins. *Biol. Bull., 142:* 489 (1972).

————, and LEVINE, E. E. Some enzymatic properties of axonemes from the cilia of *Pecten irradians. J. Cell Biol., 46:*416 (1970).

————, and LINCK, R. W. A comparison of muscle actin and ciliary microtubule protein in the mollusk *Pecten irradians. J. Mol. Biol., 40:*497 (1969).

SUMMERS, K. E., and GIBBONS, I. R. Adenosine triphosphate-induced sliding of tubules in trypsin-treated flagella of sea-urchin sperm. *Proc. Nat. Acad. Sci. U.S.A., 68:*3092 (1971).

TAMM, S. I. Ciliary motion in *Paramecium.* A SEM study. *J. Cell Biol., 55:*250 (1972).

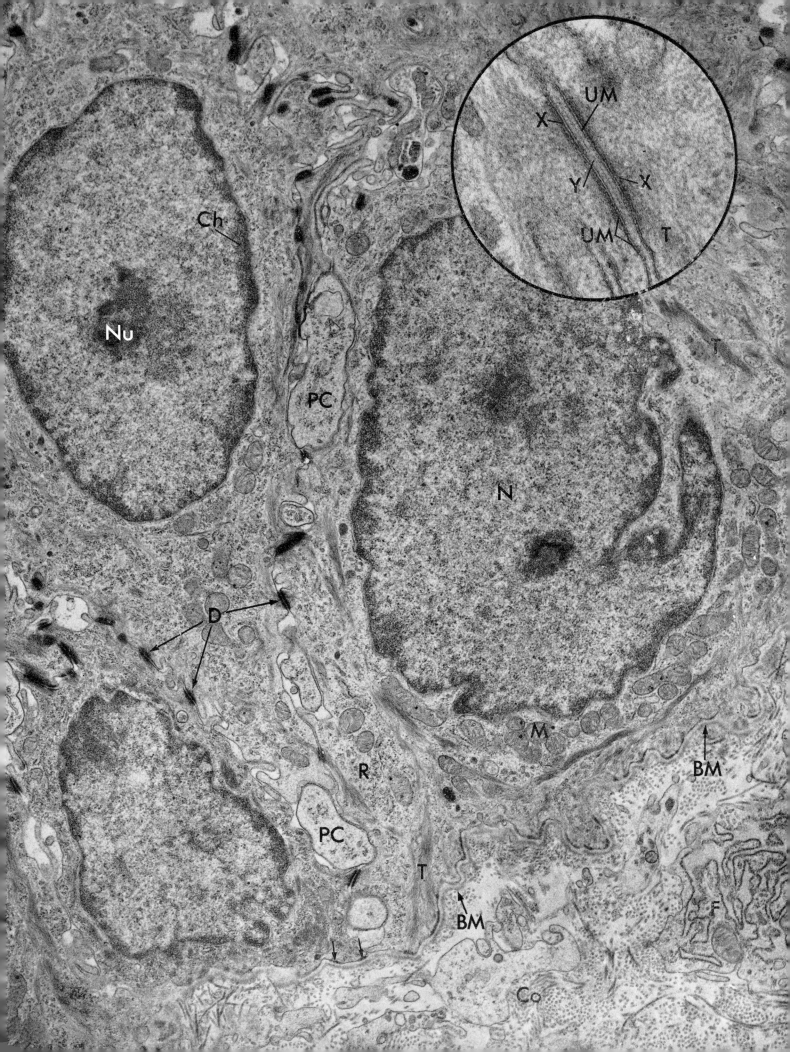

PLATE 9

The Germinal Layer of the Epidermis

The exposed body surfaces of mammals are covered by a specialized stratified squamous epithelium, the epidermis, which protects the animal from mechanical injury, invasion by foreign organisms, and loss of body fluids. The epidermis together with an underlying layer of connective tissue, the dermis, and specialized cutaneous appendages—hair, sweat, and sebaceous glands—form the mammalian skin.

The basal cell layer of the epidermis, the stratum germinativum, shown in this electron micrograph, provides cells for renewal of the epidermal population throughout life. Cells are pushed out from this layer, and only when they become part of the overlying layer, the stratum spinosum, do they begin to differentiate. As they develop, they move toward the surface of the epidermis, and by the time they have reached the uppermost layer, the stratum corneum, they have become lifeless scales, filled with horny material. Some of the features of this process, which is called keratinization, are illustrated in the text figure 9a.

Turning our attention to the germinative cells themselves, we may note that they rest on the connective tissue of the dermis, within which collagen fibers (Co) and fibroblasts (F) may be identified. A thin basal lamina (BM) separates the dermis from the cells of the stratum germinativum. The basal surfaces of the epithelial cells display irregular folds and are probably anchored to the basement membrane by the so-called "half" desmosomes (arrows). (For a description of desmosomal structure, see below.)

The nuclei (N) of the germinative cells usually have prominent nucleoli (Nu) and a thin peripheral layer of chromatin (Ch). Clusters of mitochondria (M) occur in the perinuclear cytoplasm. Strands of tonofilaments making up tonofibrils (T) are the most salient cytoplasmic feature. The fibrils, which are found generally in columnar epithelial cells, are regarded as a structural protein, giving strength to the epidermis without sacrificing permeability. Although elements of the endoplasmic reticulum are scant, free ribosomes are abundant. They are probably involved in the production of the proteins, as, for example, the tonofilaments, and later during differentiation in

the formation of keratohyalin granules (see text figure 9a) that are retained within the cells.

Continuity and integrity of the epidermis is assured by adhesion plaques, called desmosomes (D), which bind the cells together. The light microscope image of these structures led to the belief that they represented intercellular bridges, through which the cytoplasm of one cell might be continuous with the next. Now, however, electron micrographs have demonstrated that there are no open channels. Rather, as shown in the inset, dense plates (X) are aligned at opposite areas near the cell surface but within the cytoplasm of each cell. The plasma membranes that mark the limits of each cell show the trilaminar structure characteristic of the unit membrane (UM) (see Plate 2). In the region of the desmosome the innermost of the two dense lines is closely associated with the dense material of the plate (X). Usually the space between adjacent epithelial cells lacks evident content or structure, but in the region of the demosome this space is occupied by a material (Y) having, presumably, some cementing properties. As in this case the cementing material is bisected by a thin dense central lamina, seen here in cross section as a fine dense line equidistant between the two cell membranes. The intracellular side of the desmosome has attached to it tonofilaments (T), which are, in many instances, continuous with those in the fiber bundles.

The basal layer of the epidermis also contains a cell type quite different from that already described. This is the melanocyte or pigment cell, which does not begin its development within the epidermis. Instead, at an early stage in embryonic life, precursor pigment cells migrate from an area of tissue near the developing central nervous system and finally enter the epidermis. There is no difficulty involved in distinguishing these cells because they lack the tonofibrils of their neighbors and are not anchored into position by desmosomes. *Dendritic* is the adjective often applied to them, and their endings are wedged in among the more regularly arranged cells around them. Several thin processes of pigment cell cytoplasm (PC) are seen in this micrograph. Because this tissue was taken from an albino rat, the dense

43

pigment granules are not evident. In normal skin the melanoblasts or melanocytes are the only sites of pigment formation. Once formed, the pigment granules can be transferred from the melanocytes to other epidermal cells and to dermal phagocytes.

From the skin of the newborn rat
Magnification × 12,500
Inset from the esophagus of the bat
Magnification × 124,000

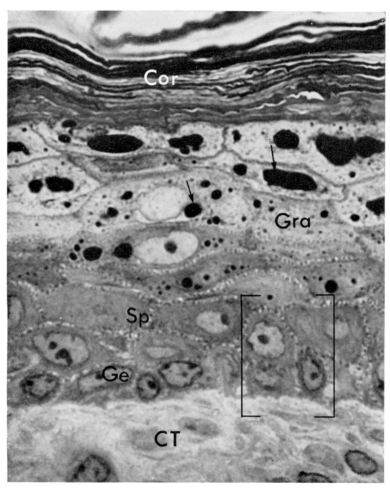

Text Figure 9a

The epidermis of a newborn rat is shown in cross section in this light micrograph, and important features of the mammalian keratinizing epithelium can be seen. An area such as that discussed in the text of Plate 9 is indicated within the brackets. It includes primarily cells of the stratum germinativum (Ge) together with some of the connective tissue (CT) of the dermis. The larger, round cells of the stratum spinosum (Sp) were derived from the basal cells and have begun to differentiate. The "prickles" on their surfaces are due to the many desmosomes that attach them to one another. As the cells are pushed toward the surface by the proliferating germinativum, differentiation continues. Dense granules (arrows) appear and increase in size as the cells near the free surface of the epidermis. Therefore in the granular layer or the stratum granulosum (Gra) production of horny material, that is, the keratinization process, begins. Eventually, in the stratum corneum (Cor) the cells become lifeless scales, filled with a tough proteinaceous material but devoid of organelles that are necessary for synthetic and metabolic activities. Finally, the horny cells at the free surface are shed as new scales are formed.

From the skin of the newborn rat
Magnification × 2,000

References

BIRBECK, M. S., BREATHNACH, A. S., and EVERALL, J. D. An electron microscope study of basal melanocytes and high-level clear cells (Langerhans cells) in vitiligo. *J. Invest. Derm., 37:*51 (1961).

BONNEVILLE, M. A. Observations on epidermal differentiation in the fetal rat. *Amer. J. Anat., 123:*147 (1968).

BRODY, I., The keratinization of epidermal cells of normal guinea pig skin as revealed by electron microscopy. *J. Ultrastruct. Res., 2:*482 (1959).

CHARLES, A., and INGRAM, J. T. Electron microscope observations of the melanocyte of the human epidermis. *J. Biophysic. and Biochem. Cytol., 6:*41 (1959).

FARBMAN, A. I. Plasma membrane changes during keratinization. *Anat. Rec., 156:*269 (1966).

FRASER,, R. D. B., MacRAE, T. P., and ROGERS, G. E. Keratins: Their Composition, Structure and Biosynthesis. Springfield, Ill., Charles C Thomas (1972).

GIROUD, A., and LEBLOND, C. P. The keratinization of epidermis and its derivatives, especially the hair, as shown by X-ray diffraction and histochemical studies. *Ann. N. Y. Acad., Sci., 53:*613 (1951).

KRAWCZYK, W. S. A pattern of epidermal cell migration during wound healing. *J. Cell Biol., 49:*247 (1971).

MARQUES-PEREIRA, J. P., and LEBLOND, C. P. Mitosis and differentiation in the stratified squamous epithelium of the rat esophagus. *Amer. J. Anat., 117:*73 (1965).

MATOLTSY, A. G. Mechanism of keratinization. *In* Fundamentals of Keratinization. E. O. Butcher and R. F. Sognnaes, editors. AAAS, Publication No. 70, Washington, D.C. (1962) p. 1.

———, and MATOLTSY, M. N. The chemical nature of keratohyalin granules of the epidermis. *J. Cell Biol., 47:*593 (1970).

———, and PARAKKAL, P. F. Membrane-coating granules of keratinizing epithelia. *J. Cell Biol., 24:*297 (1965).

MENTON, D. N., and EISEN, A. Z. Structure and organization of mammalian stratum corneum. *J. Ultrastruct. Res., 35:*247 (1971).

ODLAND, G. F. Tonofilaments and keratohyalin. *In* The Epidermis. W. Montagna and W. C. Lobitz, Jr., editors. New York, Academic Press (1964) p. 237.

OVERTON, J. Experimental manipulation of desmosome formation. *J. Cell Biol., 56:*636 (1973).

RAWLES, M. E. Origin of pigment cells from the neural crest in the mouse embryo. *Physiol. Zool., 20:*248 (1947).

RHODIN, J. A. G., and REITH, E. J. Ultrastructure of keratin in oral mucosa, skin, esophagus, claw, and hair. *In* Fundamentals of Keratinization. E. O. Butcher and R. F. Sognnaes, editors. AAAS, Publication No. 70, Washington, D.C. (1962) p. 61.

SEIJI, M., SHIMAO, K., BIRBECK, M. S. C., and FITZPATRICK, T. B. Subcellular localization of melanin biosynthesis. *Ann. N. Y. Acad. Sci., 100:*497 (1963).

ZELICKSON, A. H. Ultrastructure of Normal and Abnormal Skin. Philadelphia, Lea & Febiger (1967).

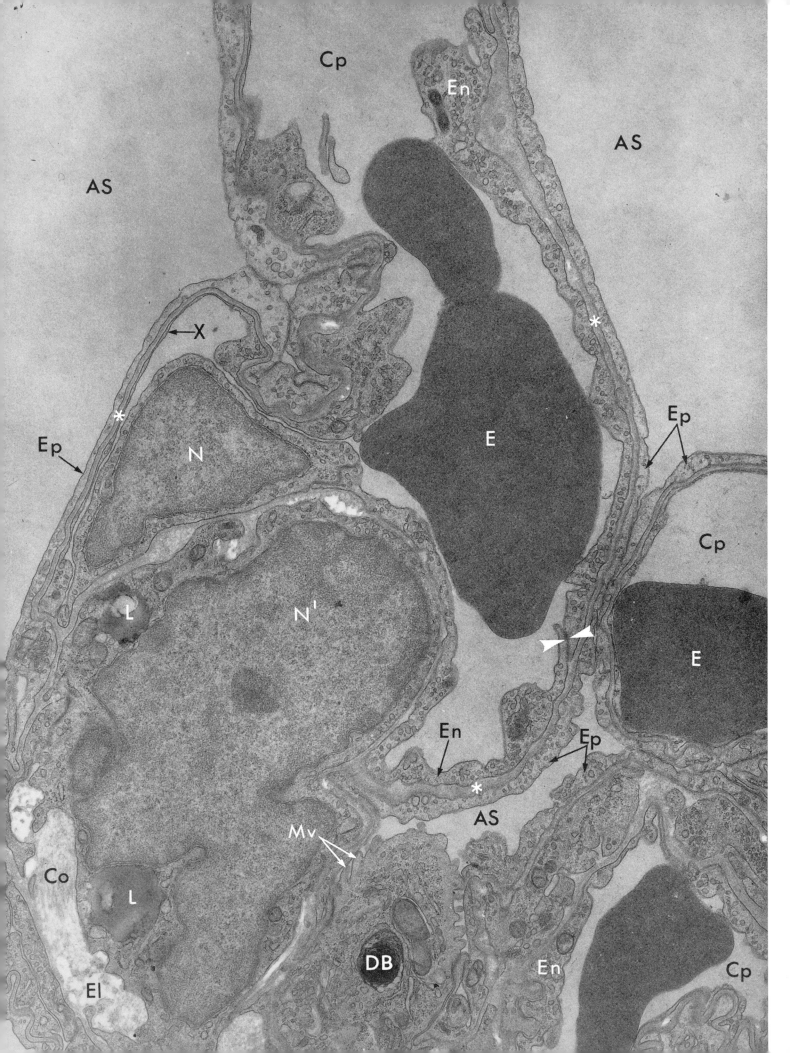

PLATE 10

The Interalveolar Septum
of the Lung and Capillary Structure

Interalveolar septa are partitions separating the alveolar sacs or air spaces of the lung. They constitute the respiratory tissue; that is, the site of gaseous exchanges between air and blood. They are so thin that it was only after examination with the electron microscope that their structure became clear. Portions of several alveolar sacs (AS), each lined by an extremely attenuated squamous epithelium (Ep), are included in this micrograph. The lining, which can be examined quite critically and over several microns in this micrograph, shows no evidence of any interruptions in its continuity.

The septum is, of course, highly vascularized. Here the thin endothelial cell linings (En) of three capillaries (Cp), each containing an erythrocyte (E), can be examined. The large irregularly shaped vessel in the upper portion of the picture has a lining that is characteristically thin and in places extremely so (X). Yet, like the alveolar epithelium, the capillary endothelium forms a complete lining. Neighboring cells are closely apposed (arrows), and no intercellular gaps of appreciable size or fenestrae within the cells themselves are found. Only in the region accommodating the nucleus (N) is the endothelial wall substantially thickened.

Each of these squamous epithelia has its own basement membrane, but where apposed to one another they seem to fuse into a common layer (∗). In such cases the resulting wall—alveolar epithelium, basement membrane, and capillary endothelium—may together be only 100 nm thick; that is, below the resolving power of the light microscope. In other areas the basement membrane of each epithelium is separated by bundles of collagen fibrils (Co) and also occasional elastic fibers (El), which play an important role in the recoil of the lung during expiration. Such fibers may be produced by septal cells, which are connective tissue elements lying among the capillaries. In this micrograph, the cell having an irregularly shaped nucleus (N′) and possessing cytoplasmic lipid droplets (L) may be of this type.

Striking features of the endothelial cells are the small membrane-bounded pits and vesicles ($\sim$50 nm in diameter), which populate the two surfaces of each cell and its intervening cytoplasm. The frequency and distribution of these structures has led some students of capillary permeability to the supposition that these structures might be involved in the transport of lipid-insoluble molecules across the endothelium. The distribution of electron opaque markers following their injection into the vascular system has in fact indicated that transport of packets (quanta) of material within vesicles may occur in the case of large molecules such as plasma proteins or antibodies. However, for the bulk transport demonstrable in physiological studies the pits and vesicles are patently inadequate, and recent investigations with the enzyme peroxidase as a marker make it clear that molecules of relatively low molecular weight pass between rather than through the endothelial cells. This conclusion is strengthened by careful examinations of contact zones between endothelial cells. Such studies reveal that in this case tight junctions occur as discrete plaques and that no continuous intercellular bandlike seal exists. (Compare the tight junction uniting intestinal epithelial cells, Plate 7.) Rather, thin slitlike passages, probably containing polysaccharide "cementing" substances, seem to exist and are in all likelihood the chief route for transport across endothelia. The endothelium of the capillaries in the brain constitutes a notable exception, however, in that in those vessels continuous tight junctions do seal off the capillary lumen so that intercellular transport of molecules is prevented.

Air drawn into the lungs reaches the alveolar sacs only after traveling through the nasal passages and bronchial tree, where it is warmed, moistened, and to some extent cleaned. Yet foreign organisms and contaminating material do penetrate into these deepest recesses of the tissue. Indeed lungs of city dwellers, for example, are commonly blackened by accumulated carbon particles that alveolar macrophages ingest and retain. These phagocytes are able to wander freely in the interstitium of the septum and to move over the surface of the epithelium lining the air sacs.

Recently it has become known that the alveolar

epithelium is coated on its free surface by a fatty substance (a surfactant), which reduces the surface tension at the interface between the epithelium and air. Theoretically, without this provision, the surface tension at this interface would be in aggregate forceful enough to collapse the alveoli. The lipid is believed to originate from the great alveolar cell, which constitutes a second cell type within the alveolar epithelium. Such cells have small microvilli (Mv) projecting into the air space and typically contain dense bodies (DB) that are believed to be precursors of the lipid coat. The intensely osmiophilic lamellae characteristic of these inclusions have been observed apparently emerging from the great alveolar cells and are thought to spread from this source over the surfaces of the entire squamous alveolar epithelium.

From the lung of the mouse
Magnification × 18,000

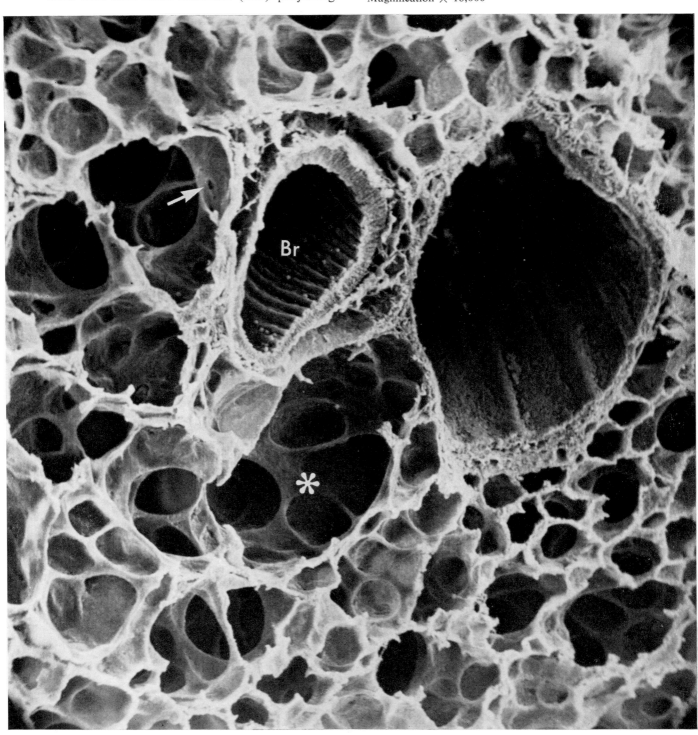

Text Figure 10a

Scanning Image of Lung

Scanning Image of Lung (Text Figure 10a)

This scanning electron micrograph surveys a field of rat lung that has been broken apart after fixation to reveal some of the air passages of the lung lobule. A bronchiole (Br), cleaved obliquely, shows its folded inner lining. This configuration of the epithelium is characteristic of the muscular ducts of the bronchiolar tree. Usually, upon fixation, the smooth muscle of the wall contracts, narrowing the lumen of the air-conducting tubules and accentuating the ridges. Identification of the larger vessel near the arteriole is uncertain, but its thin wall and large lumen may indicate that it is a venous blood vessel. A thin layer of muscle and connective tissue separates both these vessels from the respiratory tissue proper.

The rather wide passage (right center) that seems to curve through the field (*) is probably an alveolar duct, and the cavities emanating from it are alveolar sacs. The duct and the sacs constitute the bulk of the respiratory portion of the lung, the sites where exchange of gases occurs. In the septa between the sacs lies the rich vascular bed. Since this specimen was fixed by perfusion of the vascular system, red cells were removed, and the collapsed capillary vessels are not easy to identify in the torn cross-sections of the septa. Occasionally pores (arrow) are seen in the septa, connecting one alveolar sac with the next. These aid in equilibrating air within the sacs as inspiration and expiration occur. The design of the respiratory tree is such that the composition of air in the sacs is rather constant, and the partial pressures of the respiratory gases are such that uptake of oxygen and loss of carbon dioxide by the red cells in the underlying capillaries are continuously favored.

From the lung of the rat
Magnification $\times$ 200

References

BALIS, J. U., and CONEN, P. E. The role of alveolar inclusion bodies in the developing lung. *Lab. Invest., 13:* 1215 (1964).

BENNETT, H. S., LUFT, J. H., and HAMPTON, J. C. Morphological classifications of vertebrate blood capillaries. *Amer. J. Physiol., 196:*381 (1959).

BERTALANFFY, F. D. Respiratory tissue: structure, histophysiology, cytodynamics. I. Review and basic cytomorphology. *Int. Rev. Cytol., 16:*233 (1964).

BUCKINGHAM, S., HEINEMANN, H. O., SOMMERS, S. C., and MCNARY, W. F. Phospholipid synthesis in the large pulmonary alveolar cell. *Amer. J. Path., 48:*1027 (1966).

CLEMENTS, J. A. Surface tension in the lungs. *Sci. Amer. 207:*120 (December. 1962).

GREENWOOD, M. F., and HOLLAND, P. The mammalian respiratory tract surface. An SEM study. *Lab. Invest., 27:*296 (1972).

KARNOVSKY, M. J. The ultrastructural basis of capillary permeability studied with peroxidase as a tracer. *J. Cell Biol., 35:*213 (1967).

KARRER, H. E. The ultrastructure of mouse lung. General architecture of capillary and alveolar walls. *J. Biophysic. and Biochem. Cytol., 2:*241 (1956).

KISTLER, G. S., CALDWELL, P., and WEIBEL, E. R. Development of fine structural damage to alveolar and capillary lining cells in oxygen-poisoned rat lungs. *J. Cell Biol., 32:*605 (1967).

KUHN, C., III, and FINKE, E. H. The topography of the pulmonary alveolus: scanning electron microscopy using different fixations. *J. Ultrastruct. Res., 38:*161 (1972).

LOW, F. N. The pulmonary alveolar epithelium of laboratory mammals and man. *Anat. Rec., 117:*241 (1953).

LUFT, J. H. Fine structure of capillary and endocapillary layer as revealed by ruthenium red. *Fed. Proc., 25:*1771 (1966).

ROSS, R., and BORNSTEIN, P. The elastic fiber. I. The separation and partial characterization of its macromolecular components. *J. Cell Biol., 40:*366 (1969).

SOROKIN, S. P. A morphologic and cytochemical study on the great alveolar cell. *J. Histochem. Cytochem., 14:*884 (1966).

TYLER, W. S., DUNGWORTH, D. L., and NOWELL, J. A. The potential of SEM in studies of experimental and spontaneous diseases. *In* Scanning Electron Microscopy/1973. O. Johari and I. Corvin, editors. Chicago, IIT Research Institute (1973) p. 404.

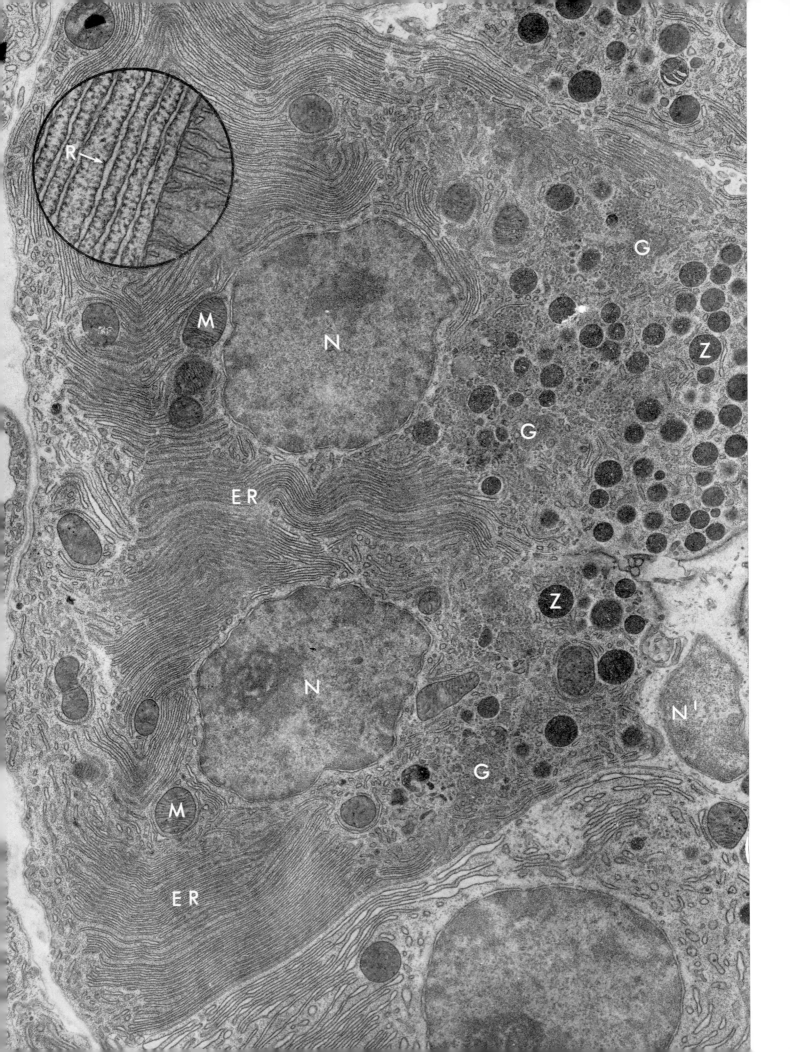

PLATE 11

Pancreatic Exocrine Cells

Secretion is one of the major activities of epithelial tissues. Toward this end the cytological machinery of glandular tissues is specialized to produce a product for export. The pancreatic exocrine cells are really a developmental derivative of the intestinal epithelium, differentiated to produce digestive enzymes, which are poured into ducts and thence reach a section of the intestine called the duodenum. The homogeneity of the tissue in terms of cell type has allowed useful cell fractionation and correlative biochemical experiments, which together have revealed roles of the cytoplasmic organelles in secretion. Thus we have come to know specific details of protein synthesis in the pancreas as well as to recognize important generalizations regarding secretory activities.

In this micrograph two pyramidal-shaped pancreatic acinar cells with centrally located nuclei (N) are shown in a fairly inactive state. Their lateral and basal cytoplasm contains relatively few mitochondria (M) and many large flattened cisternae of the endoplasmic reticulum (ER). Ribosomes in tremendous numbers (see inset, R) encrust the outer surfaces of the cisternal membranes. The ribonucleic acid of these small dense particles accounts for the well-known affinity of these cells for basic dyes. The apical ends of the cells are filled primarily by an extensive Golgi region (G) and by zymogen granules (Z), the enzyme-containing secretion droplets.

The cycle of synthesis and secretion of proteolytic enzymes is conveniently followed in these cells. After the animal has been fed, zymogen granules are discharged from the cell, and their regeneration begins anew. If one examines these cells at suitable time intervals during this period, one can observe the steps in the process and the involvement of various cytoplasmic organelles. Furthermore, if one feeds isotopically labeled amino acids, one can follow their incorporation into proteins (enzymes) and their movements as the proteins are condensed into droplets.

The labeled compounds are identified in two ways. By means of differential centrifugation the cellular organelles are segregated according to density following their release into a suitable medium upon disruption of the cell membranes. At earliest time periods the label is found asso-ciated with the "microsome" fraction, a small particulate fraction containing ribosomes and ER membranes. If now the membranes of the microsomes are destroyed with detergent, a fraction of labeled ribosomes alone is obtained. Intensive experimentation with this system has led to the concepts that the ribosomes are the sites of protein synthesis and that the newly formed enzymes are segregated within the cisternae of the endoplasmic reticulum. Subsequent to these events the radioactivity appears in the zymogen granules, and the pathway by which it reaches the droplets may be inferred from the study of electron micrographs. Small droplets resembling incomplete zymogen granules are found in the Golgi region, and the images suggest that the protein has become secondarily segregated within special smooth-surfaced vesicles and packaged into droplets in preparation for secretion. Before their release from the cell the droplets become larger, and the secretion within them becomes more concentrated.

This sequence of events has been confirmed by experiments in which autoradiographic techniques were adapted for use on electron microscope preparations. After appearing initially in association with the rough-surfaced endoplasmic reticulum (as would be expected from previous results with biochemical methods), the radioactivity becomes concentrated in the Golgi region before appearing in the zymogen granules (see also text figure 11a). In fasting animals, the granules are retained in the apical cytoplasm, but with feeding they are released, free of their membranous covering, into the ducts of the gland.

The pathway of protein synthesis, segregation, concentration, and eventual transport characteristic of the pancreatic acinar cell may be common to a wide variety of cells engaged in protein synthesis. Such variations on this theme as are displayed by other cell types active in protein synthesis involve mostly the use of other routes in the transport and release of the proteins. It is with regard to intracellular transport and to the extent and mechanics of Golgi involvement that the major questions still persist. There is, for example, accumulating evidence that fibroblasts, active in the synthesis of collagen, discharge their product directly from ER cisternae into the en-

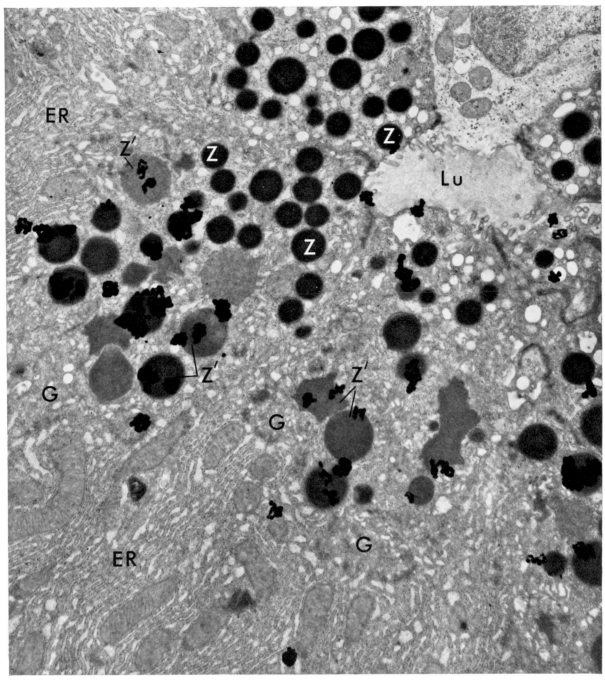

Text Figure 11a

This electron microscope autoradiograph demonstrates the localization of radioactive leucine in specific granules of pancreatic acinar cells 37 minutes after a pulse label had been administered to a slice of tissue incubated in physiological salt solution containing the labeled amino acid. Grains of reduced silver in the overlying photographic emulsion appear as irregular inklike markings near the sites from which radioactive energy was emitted. The apices of several cells bordering on the acinar lumen (Lu) each have a considerable number of secretory granules. The latter are of two types. One type, the smaller, denser round granules (Z) nearest the cell apex, consist of mature granules that are ready to be ejected into the lumen. Larger, less dense granules (Z'), which often have a serrated border, are the precursors of the former type and are called condensing vacuoles. At the moment when this tissue sample was fixed, the radioactive protein was chiefly in the condensing vacuoles. Some labeling of the more mature vacuoles had already occurred, and, if more time had been allowed between the exposure to label and fixation, these would have been the principal sites of radioactivity. It is to be noted that the rough-surfaced endoplasmic reticulum (ER), where synthesis of the zymogen proteins occurred, no longer contains any of the pulse label so prominent in it immediately after exposure. In the same way the Golgi regions (G), which are involved in transfer of the proteins from the ER into the condensing vacuoles, are not labeled, although they are heavily marked when tissue is fixed only 7 minutes after administration of label. A series of autoradiographs can thus afford a kind of time-lapse movie, revealing the progress of labeled protein through the cell and the sequential involvement of the ER, Golgi, and condensing vacuoles in the formation of the zymogen granules. We are indebted to J. D. Jamieson and G. E. Palade for this micrograph.

From the pancreas of the guinea pig (*Cavia porcellus*)
Magnification × 13,000

vironment, thus bypassing the Golgi (see Plate 28). Plasma cells may use the same route in the secretion of gamma globulins. But additional variations on the mechanism of cell secretion are recognized and are mentioned in the texts of Plates 12, 16–19, and 35.

The cells lining the ducts of the exocrine pancreas seem relatively inactive when compared to the secretory cells. In this micrograph one such cell with nucleus marked N′ lies near the apices of the glandular cells. It is evident that this cell type contains only a few organelles and no complex membranous systems.

From the pancreas of the bat (*Myotis lucifigus*)
Magnification × 13,500
Inset × 62,000

References

AMSTERDAM, A., OHAD, I., and SCHRAMM, M. Dynamic changes in the ultrastructure of the acinar cell of the rat parotid gland during the secretory cycle. *J. Cell Biol.*, *41*:753 (1969).

———, SCHRAMM, M., OHAD, I., SALOMON, Y., and SELINGER, A. Concomitant synthesis of membrane protein and exportable protein of the secretory granule in rat parotid gland. *J. Cell Biol.*, *50*:187 (1971).

BLACK, O., JR., and WEBSTER, P. D. Protein synthesis in pancreas of fasted pigeons. *J. Cell Biol.*, *57*:1 (1973).

CARO, L. G., and PALADE, G. E. Protein synthesis, storage, and discharge in the pancreatic exocrine cell. *J. Cell Biol.*, *20*:473 (1964).

FEDORKO, M. E., and HIRSCH, J. G. Cytoplasmic granule formation in myelocytes. *J. Cell Biol.*, *29*:307 (1966).

GERBER, D., DAVIES, M., and HOKIN, L. E. The effects of secretagogues on the incorporation of [2-³H] myo-inositol into lipid in cytological fractions in the pancreas of the guinea pig *in vivo*. *J. Cell Biol.*, *56*:736 (1973).

GEUZE, J. J., and POORT, C. Cell membrane resorption in the rat exocrine pancreas cell after *in vivo* stimulation of the secretion, as studied by *in vitro* incubation with extracellular space markers. *J. Cell Biol.*, *57*:159 (1973).

JAMIESON, J. D., and PALADE, G. E. Intercellular transport of secretory proteins in the pancreatic exocrine cell. I. Role of the peripheral elements of the Golgi complex. *J. Cell Biol.*, *34*:577 (1967).

———, and PALADE, G. E. Intercellular transport of secretory proteins in the pancreatic exocrine cell. II. Transport to condensing vacuoles and zymogen granules. *J. Cell Biol.*, *34*:597 (1967).

KRAMER, M. F., and POORT, C. Protein synthesis in the pancreas of the rat after stimulation of secretion. *Z. Zellforsch. Mikrosk. Anat.*, *86*:475 (1968).

PARKS, H. F. Morphological study of the extrusion of secretory materials by the parotid glands of mouse and rat. *J. Ultrastruct. Res.*, *6*:449 (1962).

ROSS, R., and BENDITT, E. P. Wound healing and collagen formation. V. Quantitative electron microscope radio-autographic observations of proline-H³ utilization by fibroblasts. *J. Cell Biol.*, *27*:83 (1965).

SIEKEVITZ, P., and PALADE, G. E. A cytochemical study on the pancreas of the guinea pig. V. *In vivo* incorporation of leucine-1-C¹⁴ into the chymotrypsinogen of various cell fractions. *J. Biophysic. and Biochem. Cytol.*, *7*:619 (1960).

SJÖSTRAND, F. S., and HANZON, V. Membrane structures of cytoplasm and mitochondria in exocrine cells of mouse pancreas as revealed by high resolution electron microscopy. *Exp. Cell Res.*, *7*:393 (1954).

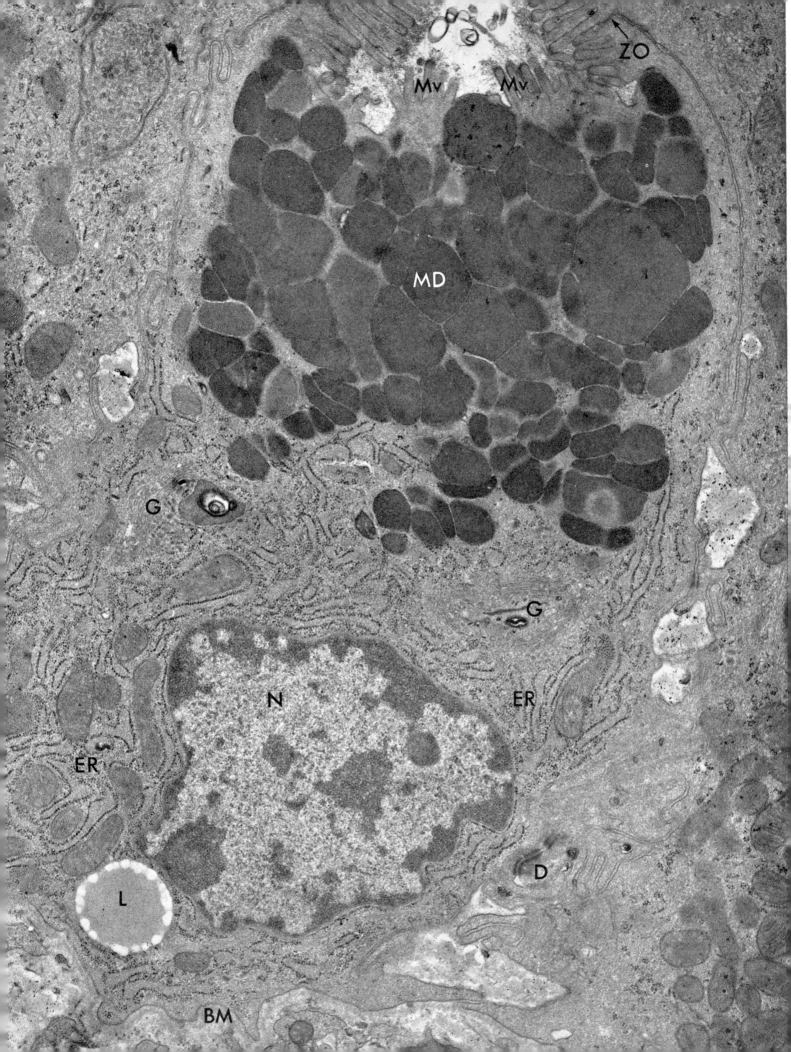

PLATE 12

The Goblet Cell

The epithelial lining of the gastrointestinal tract has among its components cells that secrete mucus, a viscous lubricating substance. Closely packed mucigen granules or mucous droplets (MD) often cause a bulging expansion of the apical cytoplasm above the narrower nucleus-containing "stem," hence the name *goblet cell*. The granules or droplets have been shown by their staining properties to contain polysaccharides, and analysis of mucous secretion has revealed it to be a complex carbohydrate, that is, a carbohydrate bound to protein, in this instance of the type referred to as glycoprotein.

The extensive knowledge of the cytological events in the production of protein secretion (see Plate 11) demands a comparison of the fine structure of the goblet cell with the pancreatic acinar cell in the hope of revealing similarities and differences in their secretory processes. In the basal cytoplasm surrounding the nucleus (N) of the columnar goblet cells, the rough-surfaced endoplasmic reticulum (ER) abounds, as one would expect in a protein secreting cell (see also Plates 1 and 2). Lipid inclusions (L) are common. Near the apical pole of the nucleus a large Golgi complex (G) underlies the mass of mucous droplets (MD). The Golgi has been described in this instance as a cup-shaped structure with its base oriented toward the nucleus. Within the cup the cavities of the Golgi saccules become distended and in the central region approach the size of smaller mucigen granules. Furthermore the contents of saccules and granules are quite similar. This observation reinforces the suggestion that the membranous Golgi saccules transform into mucous droplets.

As new droplets form, they push older ones toward the apical surface of the cell. There seems to be relatively little increase in the size of the droplets once they leave the Golgi region, and contrary to earlier electron microscope studies, recent examination of well-preserved cells indicates that the droplets are encased by a membrane during their intracellular existence and are still so enclosed as they emerge from the apical surface through gaps in the plasma membrane. In this micrograph a mucous droplet lies close to the apical membrane between two groups of microvilli (Mv). Once free in the lumen of the intestine the membrane breaks down, and the mucus is released.

Microscopic examination therefore reveals a pattern of droplet formation in many ways similar to that of protein secretion. Yet there may be essential differences between protein and mucous secretion, especially in the nature of Golgi involvement. Because goblet cells do not aggregate to form homogeneous cell populations, they are not amenable to analysis by differential centrifugation. Autoradiography, however, has proved an effective tool for studying the assembly of mucous droplets (compare Plate 11). Following the intraperitoneal injection of tritiated glucose into young rats, samples of tissue can be taken after suitable intervals. In intestinal epithelium fixed only minutes after injection, silver grains can be detected over the Golgi saccules, but the ER remains essentially unlabeled. With time, the label appears associated with mucous droplets in the Golgi "cup"; later it is found near the apical granules, and eventually it can be detected over the mucus in the intestinal lumen. These observations have led to the conclusion that the labeled glucose is incorporated into the carbohydrate component of mucus that is being synthesized within the Golgi saccules.

Chemical characterizations of the mucus presents some technical difficulties when only small amounts are available. Treatment of tissue sections with enzymes of varying specificity suggests that the cell product is a glycoprotein, since it is removed by treatment known to break down this sort of compound. When sufficient amounts of mucus can be obtained, as from the surface of the epithelium of the colon (which contains numerous goblet cells), this product, too, has been identified as glycoprotein.

Fine structural and autoradiographic studies have focused attention on the importance of the Golgi region in the production of epithelial glycoprotein. Supporting biochemical evidence from studies on liver cells indicates that the protein moiety is synthesized by the ribosomes associated with ER membranes and that the enzymes that link simple sugars to polysaccharides are to be found in a cytoplasmic membrane fraction. It seems reasonable to expect on the basis of this evidence and autoradiographic information that the

Golgi is the site where carbohydrate and protein are combined to form the glycoprotein of mucus.

The polarization and dynamic activity of the Golgi are in evidence in goblet cells. Although saccules within the cup-shaped Golgi are transformed into mucous droplets, the number of Golgi saccules does not become depleted. This indicates a constant renewal, probably at the periphery of the region. Furthermore, the production of droplets does not seem to be part of a secretory cycle, but rather is a continuous activity of the functioning goblet cell.

Goblet cells are clearly integral units of the intestinal epithelium as demonstrated by their close attachment to absorptive cells. A *zonula occludens* (ZO) is evident laterally near the apical surface, and desmosomes (D) occur at the basilateral surface. These cells have a basement membrane (BM) in common with absorptive cells of the epithelium.

From the intestine of the bat
Magnification × 23,000

References

FREEMAN, J. A. Goblet cell fine structure. *Anat. Rec., 154:*121 (1966).

HOLLMANN, K. H. The fine structure of the goblet cells in the rat intestine. *Ann. N. Y. Acad. Sci., 106:*545 (1963).

LANE, N., CARO, L., OTERO-VILARDEBÓ, L. R., and GODMAN, G. C. On the site of sulfation in colonic goblet cells. *J. Cell Biol., 21:*339 (1964).

NEUTRA, M., and LEBLOND, C. P. Synthesis of the carbohydrate of mucus in the Golgi complex as shown by electron microscope radioautography of goblet cells from rats injected with glucose-H[3]. *J. Cell Biol., 30:*119 (1966).

RAMBOURG, A., HERNANDEZ, W., and LEBLOND, C. P. Detection of complex carbohydrates in the Golgi apparatus of rat cells. *J. Cell Biol., 40:*395 (1969).

SARCIONE, E. J. The initial subcellular site of incorporation of hexoses into liver protein. *J. Biol. Chem., 239:*1686 (1964).

PLATE 13
Gastric Parietal Cells

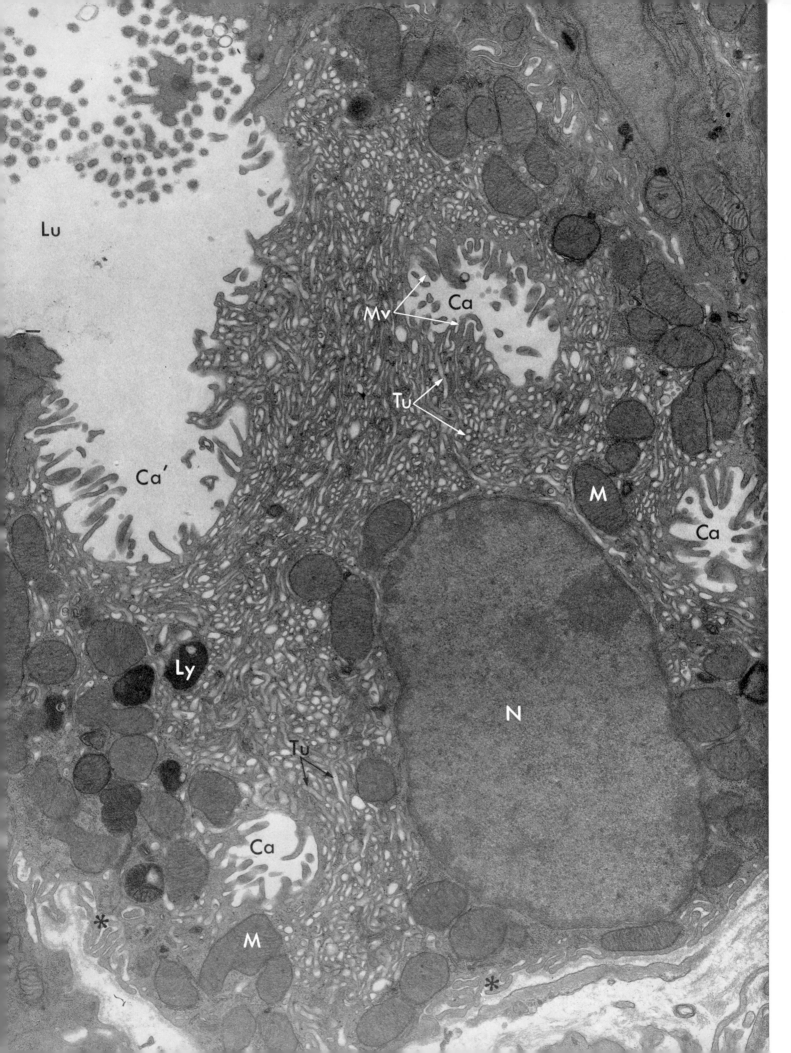

PLATE 13

Gastric Parietal Cells

Nearly one hundred years ago, the parietal cells of the gastric gland were designated as the probable sites of hydrochloric acid secretion, and this interpretation of their function continues to be generally accepted. The production of an HCl solution, approximately 0.15 N, a concentration lethal to many cells, is certainly an intriguing phenomenon, and much effort has been spent in discovering how this is accomplished. Since complete understanding still eludes us, it is not surprising that electron microscopists have been seeking clues in the fine structure of parietal cells.

A striking feature of these cells is the elaborate system of surface invaginations, or secretory canaliculi (Ca), which penetrate the cytoplasm lateral to the nucleus (N) of the cell. These passageways are connected by a common outlet (Ca') to the lumen (Lu) of the gastric gland. While light microscopists knew about this complex system, they were unaware that the canaliculi were covered with microvilli (Mv). These latter obviously increase the area of the free surface of these cells over that provided by the surface invaginations themselves. Observations on the fate of indicator substances presented to the basal surface of parietal cells have shown that, while the cytoplasm of the cells remains slightly alkaline, the canaliculi are acid. These experiments support the theory that the parietal cells produce acid, and they seem furthermore to localize the site of secretion at the free surfaces of the canaliculi.

In addition to these labyrinthine invaginations, parietal cells possess an extensive cytoplasmic membrane system consisting of tubules (Tu) limited by smooth-surfaced membranes. The tubules have been reported on occasion to open into the canaliculi. In this micrograph, the tubular system is so extensive that other cytoplasmic organelles such as mitochondria (M) and lysosomes (Ly) are crowded laterally or toward the folded basal surface (∗) of the cell. (For a discussion of lysosomes, see the texts of Plates 14 and 15.) On the assumption that such a complex system might be involved in acid secretion, comparisons have been made between cells in stomachs active and inactive with regard to this secretion. In the active state, a concentration of smooth-surfaced tubular profiles

near the canaliculi was observed. Simultaneously, the smooth-surfaced elements of the cytoplasm showed a general decrease, and the secretory canaliculi became more extensive. The inference is that the former tubules move to the surface, open to release H^+, and indeed become part of the free surface of the cell. Such observations indicate that the intracellular system of smooth-surfaced membranes performs essential functions in acid secretion.

Similar tubular cytoplasmic membrane systems apparently exist in other cells that secrete ions. In this group are included "chloride" cells that help rid certain marine fishes of excess salt. Observations on these cells have shown that the cavities of their tubular systems are continuous with the extracellular space. If, for example, electron opaque markers that cannot penetrate the plasma membrane are added to the cell's environment, they find their way into this system but are not found within the endoplasmic reticulum, the nuclear envelope, or the Golgi vesicles. These results argue that ion-secreting cells possess a unique system of plasma membrane invaginations that greatly expands the area of the cell surface. Furthermore, there seems to be no reason to suppose that the system, in spite of its superficial resemblance to the ER, is a variant of that system or that the Golgi is involved in its formation.

It is tempting to speculate on the role of this intracellular membrane system in acid secretion by the gastric parietal cell. Physiologists have discovered that while acid is secreted into the lumen of the gastric gland, an equal amount of base is formed and liberated into the blood stream. The bicarbonate ion released is believed to be formed from hydrated carbon dioxide through the mediation of the enzyme, carbonic anhydrase, which is apparently abundant in parietal cells. Just how the hydrogen ion that is secreted is formed and transported out of the apical pole of the cell is unknown, but it is known to depend on oxidative metabolic processes. It is reasonable to assume that its formation must occur at a site that is separate spatially from available hydroxyl ion and that it must be trapped in some way in order to prevent the water formation that would normally be expected. One may suppose in this regard that the system of tubules within the parietal cell could represent the separate surfaces and compartments for this segregation. There is, in fact, accumulating evidence that mucopolysaccharides reside in the cavities of similar systems in other ion-secreting cells and are capable of binding per unit weight relatively large quantities of hydrogen or other ions. As we have noted, this system is continuous with the cell surface and thus contributes to the total surface area where ion concentration apparently occurs. Upon stimulation the intracellular membrane system with its ion load can become externalized and thus achieve the secretion of hydrochloric acid.

From the stomach of the mouse
Magnification × 17,000

References

ABEL, J. H., JR., and ELLIS, R. A. Histochemical and electron microscopic observations on the salt secreting lacrymal glands of marine turtles. *Amer. J. Anat., 118*:337 (1966).

BRADFORD, N. M., and DAVIES, R. E. The site of hydrochloric acid production in stomach as determined by indicators. *Biochem. J., 46*:414 (1950).

COPELAND, D. E. A study of salt secreting cells in the brine shrimp *Artemia salina. Protoplasma, 63*:363 (1967).

DAVIES, R. E. Gastric hydrochloric acid production—the present position. *In* Metabolic Aspects of Transport Across Cell Membranes. Q. R. Murphy, editor. Madison, The University of Wisconsin Press (1957) p. 277.

DIAMOND, J. M., and BOSSERT, W. H. Functional consequences of ultrastructural geometry in "backwards" fluid-transporting epithelia. *J. Cell Biol., 37*:694 (1968).

ELLIS, R. A., and ABEL, J. H., JR. Intercellular channels in the salt-secreting glands of marine turtles. *Science, 144*: 1340 (1964).

ERNST, S. A., and ELLIS, R. A. The development of surface specialization in the secretory epithelium of the avian salt gland in response to osmotic stress. *J. Cell Biol., 40*:305 (1969).

ITO, S., and WINCHESTER, R. J. The fine structure of the gastric mucosa in the bat. *J. Cell Biol., 16*:541 (1963).

PHILPOTT, C. W. Halide localization in the teleost chloride cell and its identification by selected area electron diffraction. *Protoplasma, 60*:7 (1965).

———. The specialized surface of a cell specialized for electrolyte transport. *J. Cell Biol., 35*:104A (1967).

———. Ionic secretion by epithelial membranes. *Ciba Found. Study Group, 32*:68 (1968).

———, and COPELAND, D. E. Fine structure of chloride cells from three species of *Fundulus. J. Cell Biol., 18*: 389 (1963).

ROBERTSON, R. N. The separation of protons and electrons as a fundamental biological process. *Endeavour, 26*:134 (1967).

SEDAR, A. W. Fine structure of the stimulated oxyntic cell. *Fed. Proc., 24*:1360 (1965).

VIAL, J. D., and ORREGO, H. Electron microscope observations on the fine structure of parietal cells. *J. Biophysic. and Biochem. Cytol., 7*:367 (1960).

PLATE 14

The Hepatic Parenchymal Cell

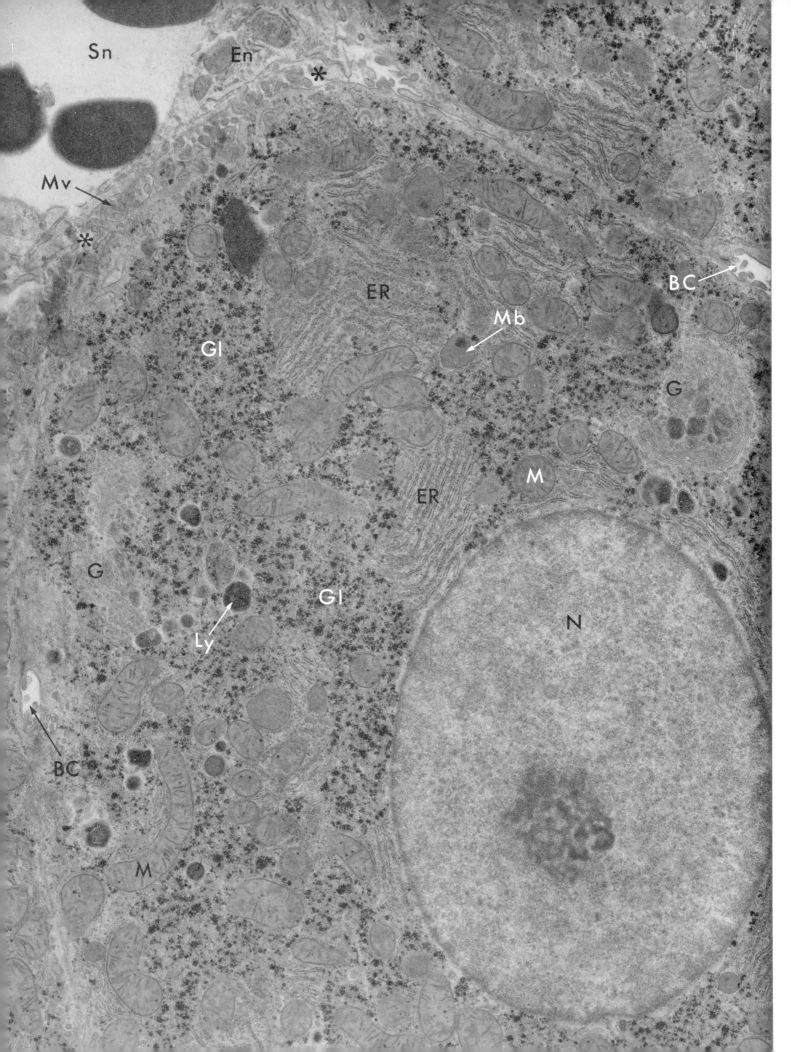

PLATE 14

The Hepatic Parenchymal Cell

The liver cell or hepatocyte is one of the most essential cells in the animal body. It is the unit of structure in a vast organ (1.5 kgs in man) which, like other glands of the gastrointestinal tract, develops from a small outpocketing of the primitive gut wall. During its growth and differentiation it takes on a variety of functions; and, though it retains only part of these in the adult, it still performs a number that are of great importance to the whole animal. It is interesting to see how the fine structure of this cell is adapted to the performance of these several roles.

Histologically, this complex epithelial tissue mingles intimately with a finely divided vascular portal system through which the blood from the intestine and spleen percolates slowly in humans but still at a total volume rate of better than one liter per minute. During its passage carbohydrates and especially glucose are removed and stored as glycogen. Amino acids are taken up for protein synthesis. One product of this synthesis, serum albumin, is manufactured in a few minutes and eventually secreted into the blood stream. Globulins, fibrinogen, and other proteins as well as lipoproteins are made in smaller amounts. Toxic agents of a wide variety, which find their way into the circulation, are removed and in most cases detoxified.

In addition to absorbing materials for storage and synthesis, the liver cell responds to circulating hormones such as epinephrine, insulin, and glucagon by secreting as an endocrine gland. For example, glucose may be mobilized from stored glycogen and put into the circulation. As an exocrine gland, the liver secretes bile salts derived from the catabolism of blood pigments. Amino acids are deaminated and thus made available to carbohydrate metabolism, while the ammonia formed is detoxified by incorporation into relatively harmless urea. These and other functions of lesser significance contribute to making the liver cell one of the most diversely active in the body.

It is not surprising to find the cytoplasm of this cell (shown here surrounding the nucleus N) morphologically one of the most complicated. No single organelle or system seems to dominate in its substructure. The rough endoplasmic reticulum, for example, does not constitute the major part of the cytoplasm as in the cells of the exocrine pancreas (Plate 11). Instead, the thin flattened cisternae (ER) are assembled into several bundles or stacks of eight to ten parallel units, which are distributed widely in the cytoplasm. In the normal, relatively quiescent cell the mitochondria (M) are located around the margins of these stacks, presumably making ATP available as a source of energy for the synthesis of plasma proteins (see Plate 2).

But these are only two of this cell's components. Marginal to these stacks of cisternae, the cytoplasm takes on a darkly speckled appearance due to dense granules of glycogen (Gl). These are regions of storage, which naturally change in appearance with time in respect to the feeding cycle. Shortly after the animal eats, they fill up with glycogen and appear very dense, whereas toward the end of the fasting period only relatively few granules remain. Scattered among these are small tubules and vesicles, whose profiles can just be detected at this magnification, representing a so-called smooth form of the ER (see Plates 7 and 15 and text figure 15a).

Like other cells, the hepatocyte possesses a Golgi complex or component (G). This tends to become localized in the cell in the vicinity of a bile canaliculus (BC) and consists, as it does in other cells (see Plates 3 and 12), of several flattened vesicles arranged in a parallel, cup-shaped array. Near its concave side there are usually juxtaposed some granule-filled and dilated vesicles (see also Plate 15).

Though the pattern is not evident in this or any single micrograph of a liver cell, examination of many pictures can demonstrate that the distribution of these membrane systems and organelles follows a pattern that brings one margin of the rough-surfaced ER stacks into apposition with glycogen-rich cytoplasm and the opposite margin close to a Golgi complex. These and other aspects of this organization are evident in the liver cell diagram (text figure 15b).

A number of dense masses, appearing round in outline and polymorphic in content, can be identified in the vicinity of the Golgi. These structures are called lysosomes (Ly) because they have been shown to contain a number of hydrolytic enzymes. Within their enclosing membranes small packets of liver cell cytoplasm are walled off and set apart for digestion. Through this device, the liver cell may destroy parts of its machinery that may have been damaged or have ceased to function normally. The products of this digestion are available to the cell for gluconeogenesis (new production of glucose) or regeneration of structure. Another type of enzyme-rich granule, called the microbody or peroxisome (Mb), is characterized by the presence within it of a dense nucleoid. A further examination and discussion of the organelles (apart from the nucleus, N) occurring in hepatic parenchymal cells will be found in the text for Plate 15.

The parenchymal cells of the liver associate to form a cuboidal epithelium that possesses some unusual characteristics. The basal surface of the tissue is exposed to the vascular supply, represented by the blood sinusoids (Sn), a special type of capillary. The usual tissue relationship would place (a) the capillary endothelium, (b) its basement membrane, (c) an interstitial connective tissue matrix, and (d) the epithelial basement membrane between the epithelial cell and the circulating tissue or blood (see Plates 6, 17, and 26). But in the case of the liver these barriers are reduced to remnants at most, so that the blood plasma may come into direct contact with the hepatocyte, thus facilitating rapid exchanges between the two. The sinusoidal endothelium (En) is discontinuous, and the pores (fenestrae) in the sieve, though not large enough to pass erythrocytes, certainly provide for free movement of plasma. Only insignificant bits of the basement membrane are evident, and the interstitium has only occasional bundles of collagen fibrils. The basal surface of the liver cell is covered by microvilli (Mv), which increase its surface area and thus constitute another specialization that aids in rapid exchange of metabolites. These short microvilli project into a space (*) called the space of Disse, which is limited on the opposite side by the sinusoidal endothelium (En). This space is continuous with the narrower ones between the cells and eventually with a system of lymph channels designed to return the extravascular plasma to the circulation. These features of liver tissue fine structure are illustrated only in part by the sinusoid (Sn) containing dense erythrocytes (upper left in this micrograph).

The apical or free surface of the hepatic epithelium is as unusual as the basal surface just described. It is represented by the relatively small surface of the liver cell that faces on and limits the bile canaliculi (BC). Only two contiguous hepatocytes are involved in forming the capillary, and their free surfaces exposed to the lumen of this tiny duct are joined at their edges by relatively tight junctions. The junctions are similar, though not identical, to those which are normally found joining intestinal epithelial cells (Plate 7). One of these attachment devices can be seen just above the arrow to the bile capillary at the right. It can be noted as well that the liver cell surfaces bordering the capillary show a few short microvilli.

From the liver of the rat
Magnification $\times$ 13,000

References

BRUNI, C., and PORTER, K. R. The fine structure of the parenchymal cell of the normal rat liver. *Amer. J. Path.,* 46:691 (1965).

DEANE, H. W. The basophilic bodies in hepatic cells. *Amer. J. Anat.,* 78:227 (1946).

DE DUVE, C. The lysosome concept. *In* Ciba Foundation Symposium on Lysosomes. A. V. S. de Reuck and M. P. Cameron, editors. London, J. and A. Churchill, Ltd. (1963) p. 1.

FAWCETT, D. W. Observations on the cytology and electron microscopy of hepatic cells. *J. Nat. Cancer Inst.,* 15:1475 (1955).

HEATH, T., and WISSIG, S. L. Fine structure of the surface of mouse hepatic cells. *Amer. J. Anat.,* 119:97 (1966).

NOVIKOFF, A. B., and ESSNER, E. The liver cell. *Amer. J. Med.,* 29:102 (1960).

PETERS, T., JR. The biosynthesis of rat serum albumin. II. Intracellular phenomena in the secretion of newly formed albumin. *J. Biol. Chem.,* 237:1186 (1962).

PLATE 15

Hepatic Cell Cytoplasm

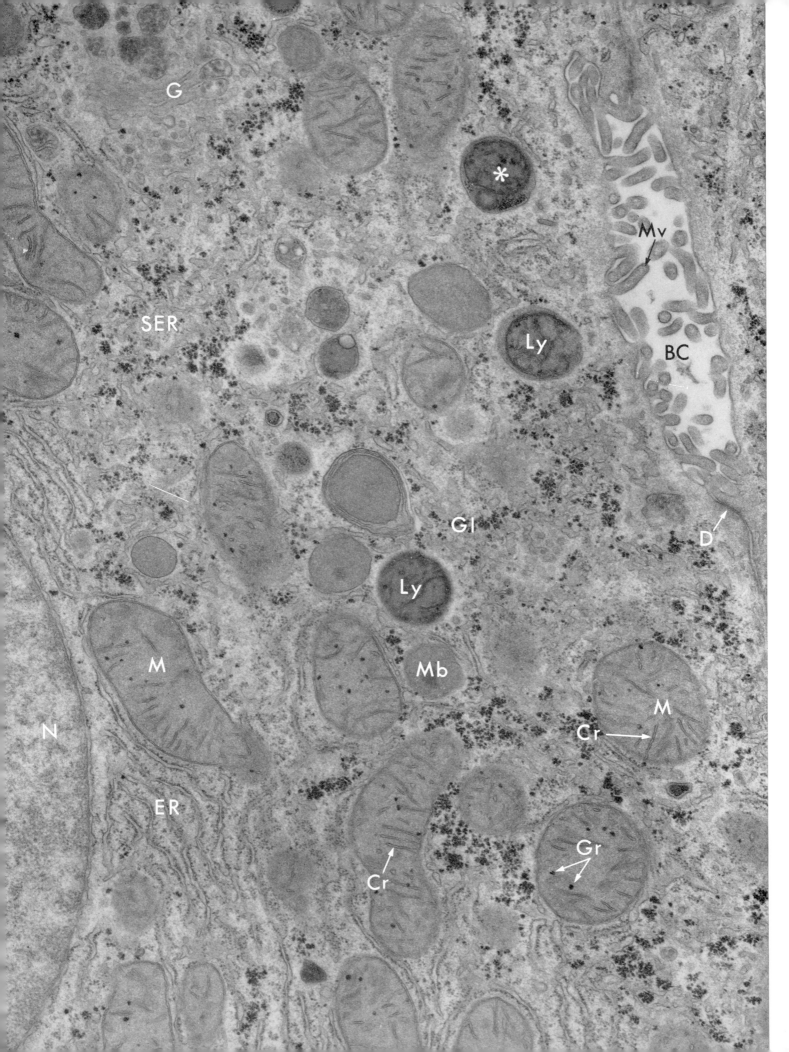

PLATE 15

Hepatic Cell Cytoplasm

Among the many cell types of the vertebrate animal none has been more extensively studied with the combined techniques of cytochemistry and electron microscopy than that of the liver. The reason resides mainly in the fact that liver as a tissue is made up very largely of one kind of cell, the hepatocyte, shown in Plate 14. It follows that any isolated cell fraction (e.g., mitochondria) does not include in important amounts contaminants from other cell types. This favorable feature of liver and the vastly improved techniques for obtaining single element fractions have contributed to the growth of a very rich literature concerning liver cell components. All of this compels one to consider separately some of their features and functions.

This electron micrograph shows part of a liver parenchymal cell or hepatocyte reproduced at a magnification sufficiently high ($\times$ 30,000) to show the fine structural details of several liver cell components. Additional details are depicted in text figures 15a and 15b.

A small part of the cell nucleus (N) is included at the lower left in the picture. It is limited by the usual two-membrane envelope, appearing quite similar in profile to the cisternae of the endoplasmic reticulum (ER) in the adjacent cytoplasm. Though not shown in this micrograph, continuities between the outer membrane of the envelope and those of the ER are frequently encountered in these cells.

In the cytoplasm of the hepatocyte two forms of this membranous reticulate system are regularly observed. In one (ER) the cisternal membranes have ribosomes on their outer surfaces. This form was discussed in the text of Plate 2, and, as mentioned there, it is the form found in cells in which protein is synthesized for secretion. Experimental studies have shown that when the amino acid leucine (radioactive) is made available to liver cells, it is, within 2 to 3 minutes, incorporated into the protein albumin. When albumin is isolated from the cell within this short period, the labeled protein is attached to microsomes derived from the rough-surfaced endoplasmic reticulum. This shows that at least for this particular protein the ribosome-rich ER is the site of synthesis.

The second form of the ER in liver cells is made up of branching tubular elements constituting a compact three-dimensional lattice. The membrane surfaces here are particle-free or smooth, hence the appellation *smooth endoplasmic reticulum* (SER). This form of the ER (better seen in text figure 15b) is continuous with the rough, and on the basis of experiments in which ^{14}C-labeled membrane components have been followed, it is known that this smooth form is derived from the rough.

In the hepatic parenchymal cell the SER shows a specific (selective) association with glycogen granules (Gl). The latter are enmeshed in the lattice spaces of the reticulum (see text figure 15b). Because of this intimate relationship and also the observed fact that the SER increases in amount during glycogen mobilization and depletion, it was proposed that the system is involved in glycogenolysis and glucose transport. In part this interpretation has been strengthened by discoveries that the SER vesicles and membranes are rich in the enzyme glucose-6-phosphatase. However, the evident mobilization of glycogen can occur in non-hepatic cells without obvious SER intervention, and the absence from the SER of phosphorylases important in the early stages of glycogen breakdown suggests now a more limited role for the SER in its liver-glycogen association. Presumably the glucose-6-phosphate as a product of glycogenolysis is picked up by the SER and dephosphorylated. The glucose then enclosed in SER vesicles is transported to the cell surface and secreted. This, at very least, is a working hypothesis.

The large membrane surfaces that the SER presents to the cytoplasmic ground substance are apparently used for other important functions of the liver cell. Some time ago it was observed that severely toxic carcinogenic agents induced a large increase in the amount of SER. The reason for this was not at once apparent but has since been clarified in discoveries that this is a general response to toxic agents. Such drugs as phenobarbital, when administered repeatedly, induce a pronounced development of the SER and simultaneously a drug-hydroxylating enzyme associated with the ER membranes. The synthesis of the detoxifying enzyme is in turn dependent on

(Continued on page 69)

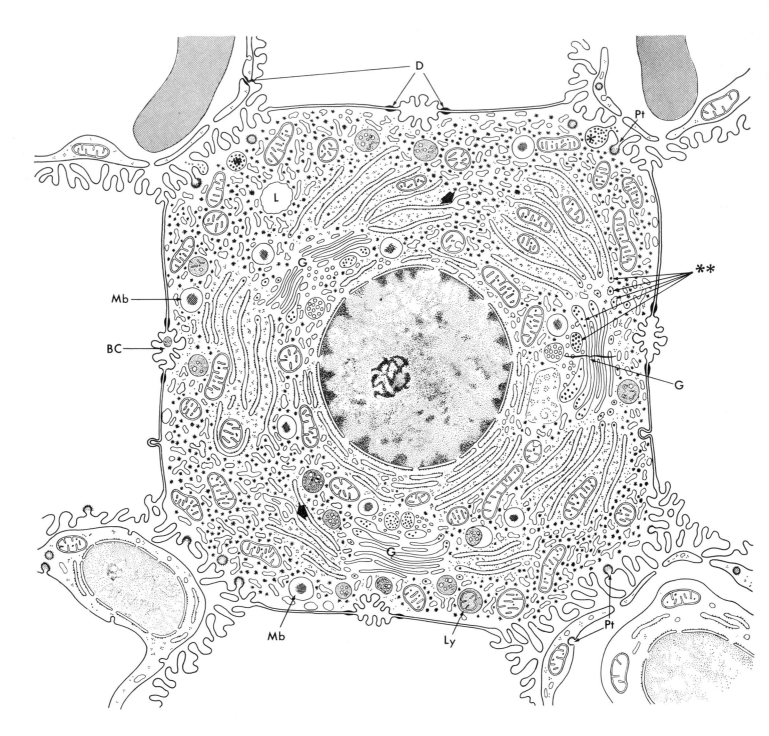

Text Figure 15a

The fine structure of a cell or tissue is frequently clarified by the construction of a diagram of the whole or its parts. Such constructed images represent distillations of the information gathered from the study of many micrographs. As such they are important for the student and even more important for the investigator, who is obliged in making the diagram to clarify his concepts of structural relationships, which might otherwise remain confused. In representing interpretations of observations, diagrams are seldom absolutely correct in all they show and

are therefore valuable only as temporary aids to understanding.

This drawing represents a single parenchymal cell of the rat liver. It is surrounded by and closely contiguous with four other hepatic cells (which are not drawn in) and four sinusoids or capillaries of the blood supply. The surfaces of the hepatocyte adjacent to the sinusoids possess many microvilli and are equivalent to the basal poles or surfaces of this epithelial cell, unusual in this instance in not being

(*Continued*)

(*Text figure 15a Continued*)
underlain by a basement membrane. The sinusoids are limited by thin endothelial cells with fenestrae, as illustrated. Red blood cells are represented in the two at the upper right and left corners of the picture; a white cell is in that at the lower right.

After examining Plates 1 through 4, 14, and 15 the student should have no difficulty in recognizing such prominent constituents of the cell as nucleus, mitochondria, and the long slender profiles of cisternae belonging to the endoplasmic reticulum or ER. Small particles or ribosomes are abundant on the surfaces of the latter and in the ground substance between them. The cisternae are ordinarily arranged in stacks of 6 to 12 units. One margin of such an assemblage frequently lies adjacent to a Golgi region. It will also be noted that small granules of a particular type occupy the ends of these cisternae and the vesicles intervening between ER and Golgi as well as the expanded ends of cisternae and spherical vesicles belonging to the Golgi proper (G). These images (**) are designed to show the mechanism of protein transport from ER to Golgi, where packaging for export from the cell takes place. The product is subsequently released from the secretion granule into the space of Disse (as shown at *). The opposite margin of the stacks of ER cisternae borders on masses of glycogen and associated vesicles belonging to the smooth endoplasmic reticulum or SER. These two forms are often continuous as though the smooth form may develop from the rough. Such continuities are indicated by thick arrows. It is thought that the smooth ER is involved in the transport of glucose from the liver cell during glycogenolysis. Other prominent components of the hepatic cell cytoplasm include lysosomes (Ly), containing remnants of organelles and ground substance apparently set aside for hydrolysis, and microbodies or peroxisomes (Mb). A single lipid granule is indicated at L.

Various differentiations of the cell surface are shown, though their full functional meaning is not completely clarified as yet. The surface adjacent to and limiting the bile canaliculus (BC) is increased in area by many microvilli. This represents the free surface of this cell, and as is common in epithelial cells it is limited by close junctions and desmosomes (D). Besides showing numerous microvilli on its sinusoidal surface, the liver cell possesses peculiar pits (Pt) or spherical depressions with what appear as short bristles on their cytoplasmic surfaces. These same structures are present on the endothelial cells and Kupffer cells lining the sinusoids. They are thought to be involved in the selective uptake of proteins and possibly other macromolecules from the circulating blood plasma.

(Text Continued)

the production of a rather short-lived messenger RNA (see Plate 4). Thus the liver cell has evolved a remarkable device for the rapid detoxification of otherwise destructive foreign agents that find their way into the body.

The Golgi complex (G) of the hepatocyte is similar in its major characteristics to that in other cells (see Plates 3 and 12). In this micrograph it appears as a stack of 4 or 5 flattened cisternae limited by smooth-surfaced membranes. That the outlines of these are not more distinctly shown here is due to the somewhat oblique orientation of the section relative to the vertical axis of the stack. At their margins it is common for the Golgi cisternae to be inflated and to contain small dense granules about 500 A in diameter and currently interpreted as lipoprotein. The presence of similar granules in vesicles of the Golgi-associated ER and on the concave side of the Golgi complex in large spherical vesicles (see top left in micrograph) has been taken to mean that ER-sequestered materials are fed into one pole of the Golgi region to emerge from the other face packaged (and possibly refined) into larger quantities for transport thence to the cell surface. This is only one of several activities that have been postulated for the hepatocyte Golgi. For it, as for the others, the information yet available is meager and largely morphological.

The dense bodies (Ly) usually located near the surfaces of the hepatocyte bordering the bile canaliculi are representative of lysosomes. It is characteristic for them to contain recognizable portions of cytoplasmic components (*), such as glycogen, mitochondria, or cisternae of the ER. These contents show in turn varying degrees of disintegration, and this morphological picture fits the well-established functional concept that hydrolytic enzymes (phosphatases and proteases) are active within these tiny vesicles. Experiments have demonstrated that in starvation or under other pathological conditions cells possess a mechanism for self-consumption. Small packets of cytoplasm are walled off and digested. Apparently the products of digestion are in part reassimilated. The vesicles involved in this kind of controlled autolysis have been variously called cytolysosomes or autophagic vesicles. Remnants of digestion have been found in the bile canaliculi, suggesting that lysosomes discharge their residual contents from the cell and animal by this route.

Other granular components of the liver cell are reminiscent of secretory granules found in other cell types, such as eosinophils (Plate 34). In the hepatocyte they are called microbodies or peroxi-

(Continued on page 71)

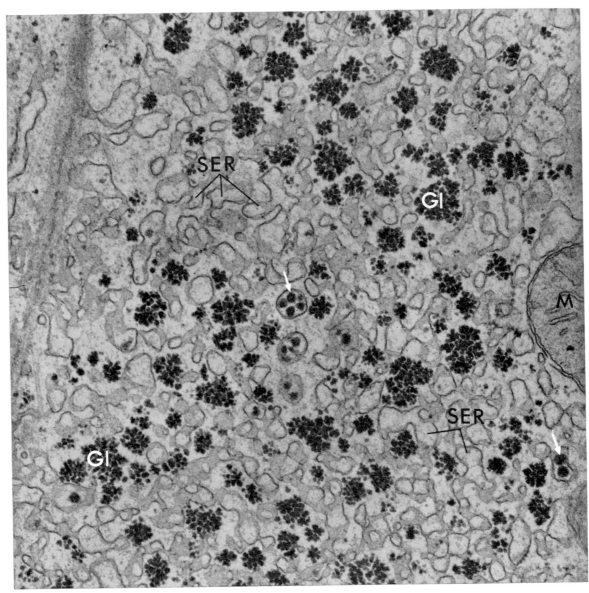

Text Figure 15b

This superb micrograph depicts the form of the smooth-surfaced endoplasmic reticulum in a hepatocyte and its relation to glycogen. The profiles of vesicles and tubules (SER) caught in this thin section show the system to be a tridimensional lattice. The spaces between the tubules and vesicles contain the dense glycogen particles (Gl) and make the relationship between ER and glycogen a very intimate one. Obviously, exchanges between the smooth ER with its large membrane surfaces and products of glycogenolysis would be easily achieved by this close association.

The small granules within some of the vesicles (arrows) are similar in size and density to granules of lipoprotein noted in Plate 15. A mitochondrion (M) enters the picture at the right, and a tangential section through the cell surface is included at the upper left.

We are indebted to G. Dallner, P. Siekevitz, and G. E. Palade for the micrograph, which is reprinted from the *J. Cell Biol., 30:*73 (1966).

From the liver of a three-day-old rat
Magnification × 48,000

somes (Mb). It may be observed from studying them in this micrograph that they are smaller than mitochondria (M) and that they are limited by a single membrane. Their contents consist of a homogeneous matrix enclosing a nucleoid, which at higher resolutions would appear crystalline. Furthermore, cytochemical techniques applied to the isolation and study of peroxisomes have shown them to be rich in uricase, catalase, and D-amino acid oxidase. Just where the peroxisomes arise is still in doubt, but their content of pure enzyme would implicate the rough ER in their synthesis. It is generally thought that they fuse with or otherwise become part of autophagic vesicles.

Other details of hepatocyte structure may be noted in this micrograph. The bile canaliculus (BC), which is formed by the apical surfaces of two liver cells, is sealed off from intercellular spaces by close junctions, which were earlier interpreted as desmosomes (D). Microvilli (Mv) extend into the lumen of this small duct. Liver cell mitochondria (M), so often isolated and examined, are characterized by a dense matrix, relatively few cristae (Cr), and a goodly number of mitochondrial granules (Gr).

From the liver of the rat
Magnification $\times$ 30,000

References

ARSTILA, A. U., and TRUMP, B. F. Studies on cellular autophagocytosis. The formation of autophagic vacuoles in the liver after glucagon administration. *Amer. J. Pathol., 53*:687 (1968).

ASHFORD, T. P., and PORTER, K. R. Cytoplasmic components in hepatic cell lysosomes. *J. Cell Biol., 12*:198 (1962).

ASHLEY, C. A., and PETERS, T., JR. Electron microscopic radioautographic detection of sites of protein synthesis and migration in liver. *J. Cell Biol., 43*:237 (1969).

ASHWORTH, C. T., LEONARD, J. S., EIGENBRODT, E. H., and WRIGHTSMAN, F. J. Hepatic intracellular osmiophilic droplets. Effect of lipid solvents during tissue preparation. *J. Cell Biol., 31*:301 (1966).

BAUDHUIN, P. Liver peroxisomes, cytology and function. *Ann. N. Y. Acad. Sci., 168*:214 (1969).

————, BEAUFAY, H., and DE DUVE, C. Combined biochemical and morphological study of particulate fractions from rat liver. *J. Cell Biol., 26*:219 (1965).

BENNETT, G., and LEBLOND, C. P. Passage of fucose-^{3}H label from the Golgi apparatus into dense and multivesicular bodies in the duodenal columnar cells and hepatocytes of the rat. *J. Cell Biol., 51*:875 (1971).

BOLENDER, R. P., and WEIBEL, E. R. A morphometric study of the removal of phenobarbital-induced membranes from hepatocytes after cessation of treatment. *J. Cell Biol., 56*:746 (1973).

COIMBRA, A., and LEBLOND, C. P. Sites of glycogen synthesis in rat liver cells as shown by electron microscope radioautography after administration of glucose-H^3. *J. Cell Biol., 30*:151 (1966).

DALLNER, G., SIEKEVITZ, P., and PALADE, G. E. Biogenesis of endoplasmic reticulum membranes. II. Synthesis of constitutive microsomal enzymes in developing rat hepatocyte. *J. Cell Biol., 30*:97 (1966).

DE DUVE, C. The lysosome. *Sci. Amer., 208*:64 (May, 1963).

DROCHMANS, P. Morphologie du glycogène. *J. Ultrastruct. Res., 6*:141 (1962).

GLAUMAN, H. Studies on the synthesis and transport of albumin in microsomal subfractions from rat liver. *Biochim. Biophys. Acta, 224*:206 (1970).

HOLT, S. J., and HICKS, R. M. The localization of acid phosphatase in rat liver cells as revealed by combined cytochemical staining and electron microscopy. *J. Biophysic. and Biochem. Cystol., 11*:47 (1961).

HRUBAN, Z., and SWIFT, H. Uricase: localization in hepatic microbodies. *Science, 146*:1316 (1964).

JONES, A. L., and FAWCETT, D. W. Hypertrophy of the agranular endoplasmic reticulum in hamster liver induced by phenobarbital. *J. Histochem. Cytochem., 14*:215 (1966).

JONES, A. L., RUDERMAN, N. B., and HERRERA, M. G. An electron microscopic study of lipoprotein production and release by isolated perfused rat liver. *Proc. Soc. Exp. Biol. Med., 123*:4 (1966).

KURIYAMA, Y., OMURA, I., SIEKEVITZ, P., and PALADE, G. E. Effects of phenobarbital on the synthesis and degradation of the protein components of rat liver microsomal membranes. *J. Biol. Chem., 244*:2017 (1969).

MILLONIG, G., and PORTER, K. R. Structural elements of rat liver cells involved in glycogen metabolism. In Proceedings of the European Regional Conference on Electron Microscopy, Delft, 1960. A. L. Houwink and B. J. Spit, editors. Delft, De Nederlandse Vereniging voor Electronenmicroscopie (1960) p. 655.

NILSSON, R., PETERSON, E., and DALLNER, G. Permeability of microsomal membranes isolated from rat liver. *J. Cell Biol., 56*:762 (1973).

NOVIKOFF, A. B., ESSNER, E., and QUINTANA, N. Golgi apparatus and lysosomes. *Fed. Proc., 23*:1010 (1964).

OMURA, I., SIEKEVITZ, P., and PALADE, G. E. Turnover of constituents of the endoplasmic reticulum membranes of rat hepatocytes. *J. Biol. Chem., 242*:2389 (1967).

ORRENIUS, S., ERICSSON, J. L. E., and ERNSTER, L. Phenobarbital-induced synthesis of the microsomal drug-metabolizing enzyme system and its relationship to the proliferation of endoplasmic membranes. A morphological and biochemical study. *J. Cell Biol., 25* (no. 3, part 1):627 (1965).

PORTER, K. R., and BRUNI, C. An electron microscope study of the early effects of 3'-Me-DAB on rat liver cells. *Cancer Res., 19*:997 (1959).

ROSEN, S. I., KELLY, G. W., and PETERS, V. B. Glucose-6-phosphatase in tubular endoplasmic reticulum of hepatocytes. *Science, 152*:352 (1966).

SHNITKA, T. K. Comparative ultrastructure of hepatic microbodies in some mammals and birds in relation to species differences in uricase activity. *J. Ultrastruct. Res., 16*:598 (1966).

SWIFT, H., and HRUBAN, Z. Focal degradation as a biological process. *Fed. Proc., 23*:1026 (1964).

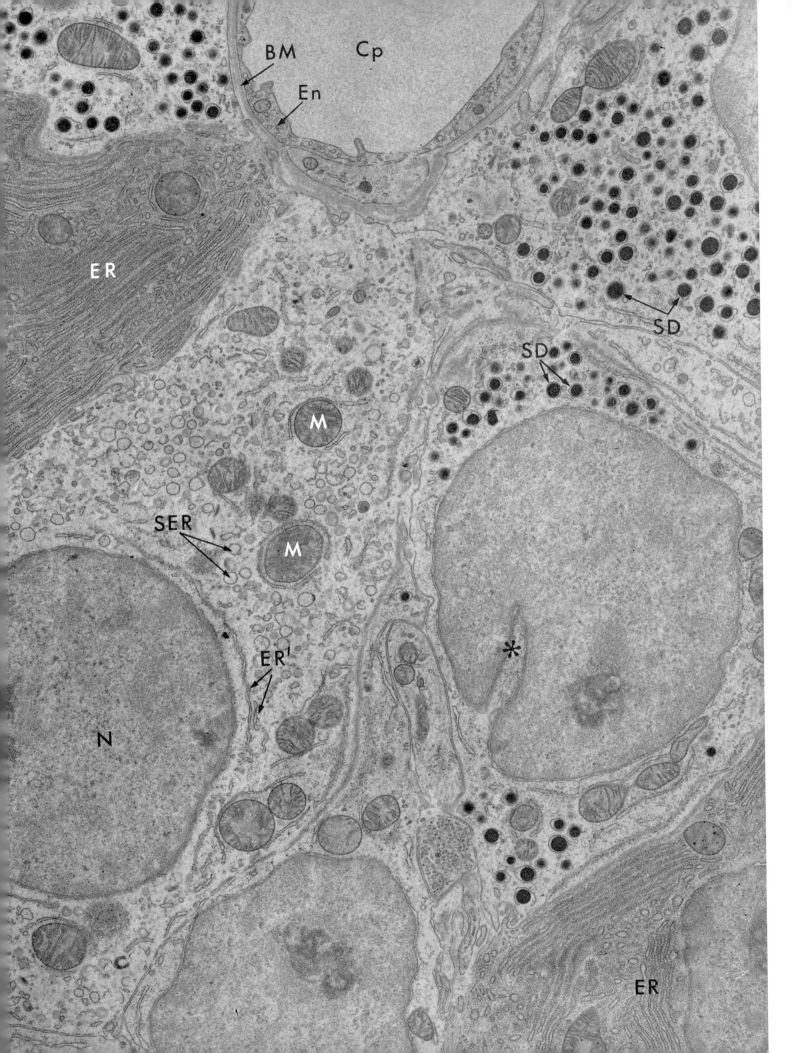

PLATE 16

Pancreatic Endocrine Cells

The pancreatic endocrine tissue, called the islets of Langerhans, differs markedly in fine structure from the exocrine tissue of that gland (see Plate 11). In this micrograph portions of two acinar cells, dense with intracellular membrane systems (ER) and associated ribosomes, may be compared with light-staining cytoplasm of the islet tissue. Characteristically, as here, endocrine cells are always close to blood capillaries (Cp) that carry away their secretory products. The endothelium (En) of this vascular tissue forms a thin but complete cell layer. It rests on a basement membrane (BM) that always constitutes part of the barrier between blood and endocrine tissue.

Several cell types are found in the islets of Langerhans, and differences in morphology suggest that they differ in function as well. The much-studied hormone, insulin, is now known to be a product of the beta cells, which have numerous dense, membrane-bounded secretory granules throughout their cytoplasm. The portions of three granule-containing cells seen in this micrograph—one with a nuclear cleft (*)—probably belong to beta cells. To be certain of this identification one should prepare an adjacent serial section for light microscopy and determine whether the granules stain with aldehyde fuchsin. This dye is believed to combine with the sulfide groups of the insulin molecule. In electron micrographs, beta cell granules (SD) are usually separated from the membranes that enclose them by a clear halo, supposed to be an artifact.

There is abundant evidence now that beta cells are the source of insulin. Originally this insight resulted from cytological studies of pancreatic islets from humans or experimental animals suffering from diabetes. Although three types of islet cells can be differentiated by their tinctorial properties, of these three it is the beta cell that degenerates in animals with diabetes and exhibits degranulation when exposed to substances that bring about insulin release. The availability of purified insulin at present allows preparation of fluorescent-labeled antibody that reacts specifically with the hormone. The fluorescent label may be detected within beta cells stained with such antibody preparations.

When insulin-producing cells are filled with cytoplasmic secretion droplets (SD), elements of the rough-surfaced endoplasmic reticulum and free ribosomes are reduced in number. When, however, the granules are few, as they are after active secretion, the rough-surfaced ER and free ribosomes are proportionally increased, presumably in response to a stimulus bringing about increased protein synthesis. The morphological events occurring during the formation of secretory droplets are not entirely clear. While some observers believe that the Golgi complex may be involved in formation of the droplets as it is in the exocrine pancreas (see Plate 11), others suggest that the droplets arise solely within the endoplasmic reticulum. The secretory product would, in this second interpretation, be assembled into granules within the cisternae of the ER, and from these, vesicles enclosing the dense granules would pinch off after first becoming free of ribosomes.

Secretion of insulin from beta cells is believed to be initiated by entrance of sodium ion into the cells, but release of membrane-free granules into the extracellular space has been difficult to detect. Nevertheless, in well-fixed specimens, evidence exists that the membrane surrounding the granule fuses with the plasma membrane and then opens to allow escape of the secretory droplet into the extracellular space around the capillaries. The secretory products would then be free to diffuse out of the cell and into the circulatory system.

Alpha cells are a second type of secretory cell occurring in the islets of Langerhans. Like the beta cells, they contain within their cytoplasm numerous dense secretory granules enclosed by membranes. Although the two types generally resemble each other in their cytoarchitecture, a distinction may be made on the basis of subtle fine structural differences. Alpha cells correspond to those cells that in preparations for the light microscope are stained positively by the phosphotungstic acid-hematoxylin method. It is believed that alpha cells secrete glucagon, the glycogenolytic hormone. The production of the small protein molecule, insulin, and the polypeptide, glucagon, are therefore believed to occur in different cells. It is interesting to note in contrast that several different enzymes are produced by a single type of pancreatic acinar cell (see Plate 11).

Cells that lack cytoplasmic secretion droplets are also present in the pancreatic islets. These have often been called chromophobic or by other names such as c cells, and one appears at the lower left in this micrograph. In the cytoplasm surrounding the spherical nucleus (N) there are a few rough-surfaced cisternae of the endoplasmic reticulum (ER′), a considerable number of smooth-walled vesicles (SER), also parts of the endoplasmic reticulum, and globular mitochondria (M). The function of this type of cell is unknown. As yet no one can say whether it represents a completely different cell type or illustrates simply one stage in the secretory cycle of either of the granule-producing cells.

From the pancreas of the bat
Magnification × 10,500

References

ANDERSSON, A. Monolayer culture of pancreatic islet cells. *In* The Structure and Metabolism of the Pancreatic Islets. S. Falkmer, B. Hellman, and I. B. Täljedal, editors. Oxford, Pergamon Press (1970) p. 73.

CARAMIA, F., MUNGER, B. L., and LACY, P. E. The ultrastructural basis for the identification of cell types in the pancreatic islets. I. Guinea pig. *Z. Zellforsch. Mikrosk. Anat., 67:*533 (1965).

HELLERSTROM C., HOWELL, S. L., EDWARDS, J. C., and ANDERSSON, A. An investigation of glucagon biosynthesis in isolated pancreatic islets of guinea pigs. *FEBS (Fed. Eur. Biochem. Soc.) Lett., 27:*97 (1972).

LACY, P. E. Electron microscopy of the beta cell of the pancreas. *Amer. J. Med., 31:*851 (1961).

LAZAROW, A. Cell types of the islets of Langerhans and the hormones they produce. *Diabetes, 6:*222 (1957).

LEVER, J. D., and FINDLAY, J. A. Similar structural bases for the storage and release of secretory material in adreno-medullary and β pancreatic cells. *Z. Zellforsch., 74:*317 (1966).

MILNER, R. D. G., and HALES, C. N. The sodium pump and insulin secretion. *Biochim. Biophys. Acta, 135:*375 (1967).

MUNGER, B. L. The secretory cycle of the pancreatic islet α-cell *Lab. Invest., 11:*885 (1962).

OPIE, E. L. Cytology of the pancreas. *In* Special Cytology, second edition. E. V. Cowdry, editor. New York, Paul B. Hoeber, Inc., vol. I (1932) p. 375.

SATO, T., HERMAN, L., and FITZGERALD, P. J. The comparative ultra-structure of the pancreatic islet of Langerhans. *Gen. Comp. Endocr., 7:*132 (1966).

WILLIAMSON, J. R., LACY, P. E., and GRISHAM, J. W. Ultrastructural changes in islets of the rat produced by tolbutamide. *Diabetes, 10:*460 (1961).

PLATE 17

The Anterior Lobe of the Pituitary:
Somatotrophs and Gonadotrophs

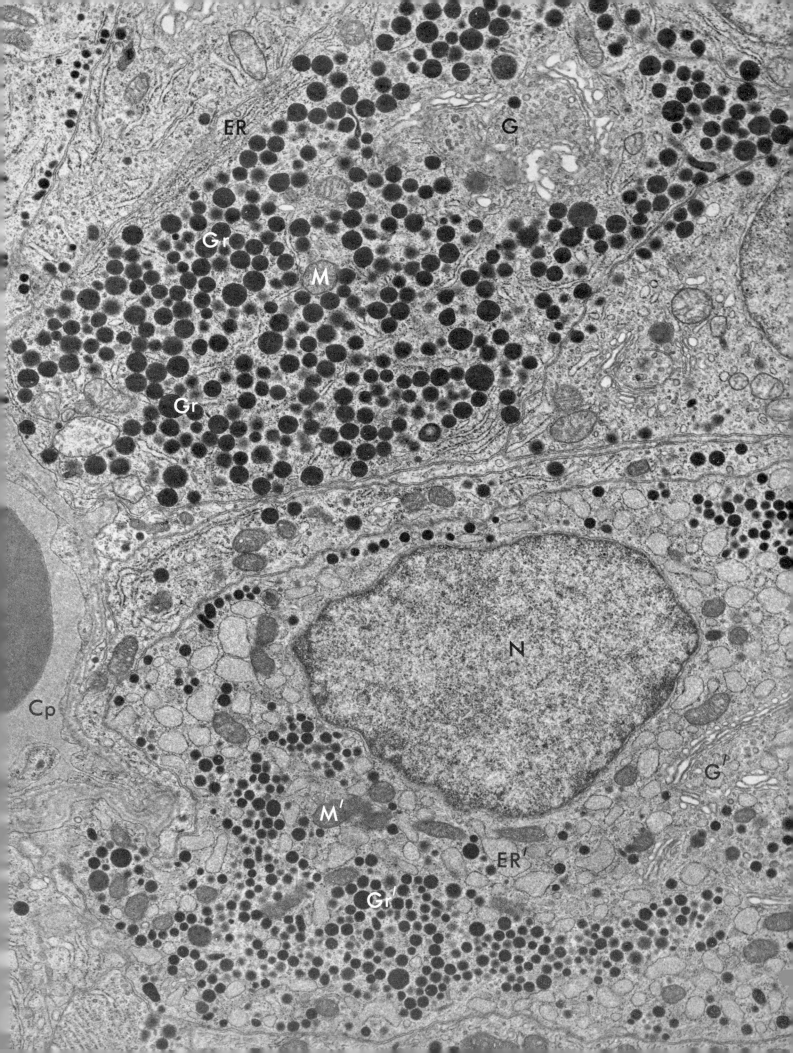

PLATE 17

The Anterior Lobe of the Pituitary: Somatotrophs and Gonadotrophs

Complexity of origin, structure, and function is characteristic of the pituitary gland, and chief among the puzzles to be solved has been the identification of the cell types that produce the multiplicity of hormones discovered through classical physiological experiments. The problem is illustrated by this micrograph. Six hormones are known to arise in the pars distalis or anterior lobe of the adenohypophysis, that portion of the pituitary arising during embryonic life from an outpocketing of the roof of the oral cavity. Distinction of cell types by their staining properties seemed at first to provide only three sorts of cells, named in regard to their affinities for dyes, i.e., acidophils, basophils, and chromophobes. The acidophils, which make up 30 to 45 per cent of the cells of the anterior lobe of the rat, form proteinaceous secretion droplets large enough to be resolved easily in the light microscope. The majority of acidophils produce growth hormone—somatotrophic hormone or STH—and are therefore called somatotrophs. One of these granule-filled somatotrophs may be seen in profile in the upper part of this micrograph. The second less numerous acidophil secretes prolactin, which promotes lactation.

Basophils, which produce glycoprotein secretions (PAS positive), include the thyrotrophs, the producers of thyroid stimulating hormone or TSH, and the gonadotrophs, which produce FSH or follicle-stimulating hormone (an activator of spermatogenesis in the male) and LH, luteinizing hormone, which is also called ICSH or interstitial cell-stimulating hormone in the male. The profile of a gonadotroph appears in the lower part of this micrograph (nucleus, N).

The cell type responsible for production of adrenocorticotrophic hormone or ACTH has not yet been positively identified, but basophils and chromophobes have been implicated as the source of this hormone. In electron micrographs the chromophobe appears as a cell with a very irregular outline and few secretory granules.

Nature performed the earliest experiments that pinpointed the acidophils as the site of growth hormone production. By 1903, acromegaly, the condition characterized by distorted overgrowth of the skeleton, was associated with the presence of pituitary tumors made up of acidophils. In subsequent investigations, extracts made from the lateral regions of the bovine adenohypophysis—areas rich in acidophils—were found superior in growth promoting properties to extracts made from median regions of the gland where acidophils are less numerous. When the pituitaries of dwarf mice were examined, acidophils were almost entirely lacking. Confirmation that the source of growth hormone has been correctly identified has come from the use of fluorescent labeled antibodies to growth hormone used as an antigen. When used to stain anterior pituitary glands, the antibody is, as one would expect, localized in the cells previously designated as somatotrophs.

Somatotrophs are easily identified in electron micrographs because they contain many dense cytoplasmic granules (Gr) that are of fairly uniform size and have a maximum diameter of about 350 mμ. Isolation of protein-containing granules of this size by means of differential centrifugation has shown that they are rich in growth hormone. This evidence therefore supports the reasonable assumption that the granules are secretion droplets.

Cisternae of rough-surfaced endoplasmic reticulum (ER) are present in moderate numbers as are mitochondria (M). Under experimental conditions, however, in which the somatotrophs are stimulated, the ER increases in prominence. A large Golgi complex (G) is a constant feature of these cells. Certain of the secretory granules found within the Golgi area are clearly surrounded by a membrane (see also text figure 17a). The formation of protein secretory granules in the pituitary is believed to involve synthesis by ribosomes of the rough-surfaced ER and assembly of the product in the Golgi complex. After assembly, the secretory granules move out of the Golgi area and may be found throughout the cytoplasm.

Release of secretory product from the cell is difficult to detect morphologically. However, in text figure 17a, secretory product lying outside the cell and free of its membranous covering may

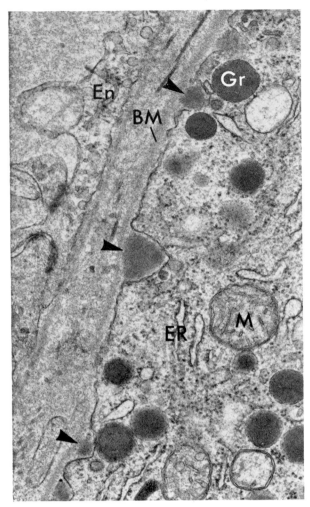

Text Figure 17a

Profiles of mitochondria (M), elements of the endoplasmic reticulum (ER), and secretion droplets (Gr) as present in the peripheral cytoplasm of a somatotroph are depicted in this micrograph. Secretory product (arrows), free of membrane covering, lies extracellularly but within small cavities in the cell surface. The extruded material has not yet penetrated the basement membrane (BM) that underlies the glandular epithelium. The endothelium (En) of an adjacent capillary, together with its basement membrane, is separated from the somatotroph by a thin layer of connective tissue. This micrograph was generously provided by Dr. Robert Cardell.

From the pituitary of the rat
Magnification × 30,000

be identified within small concavities of the plasma membrane of a somatotroph. One can easily postulate that a process of "reverse pinocytosis" could place the contents of the granule outside the cell. The extracellular material always

seems less dense than that in the intracellular granules. Furthermore, it can be detected only between the plasma membrane of the somatotroph and its closely adhering basal lamina. As soon as it leaves this area, its identity as a granule ceases, and the microscopist cannot detect the hormone after it enters the connective tissue that separates the secretory cell from nearby capillaries (Cp).

Identification of the other prominent group of cells of the anterior lobe, the gonadotrophs, in EM preparations has depended on comparison of tissues from normal and castrate animals. Castration results in accumulation of gonadotrophic hormones in the pituitary. Histologically, an increase in basophilic granules and an intensification of PAS staining (for glycoprotein) results. Careful observation has further shown that two types of gonadotrophs may be distinguished. In one, which occurs primarily at the periphery of the adenohypophysis, the response to castration is prompt, while in the second type, located in the central region of the gland, a response is noted only several weeks after castration. Histological studies of normal glands in correlation with the physiological state of the reproductive organs in the female rat reaching maturity have led to the belief that the first type, the gonadotroph lying in the peripheral region of the gland, secretes follicle-stimulating hormone or FSH and that the second type, the one slow to respond to castration, is responsible for the production of luteinizing hormone or LH.

The gonadotroph seen in this micrograph displays several features characteristic of the cell that produces FSH. Mitochondria (M') containing a rather dense matrix material are scattered among distended vesicular elements of the endoplasmic reticulum (ER') with associated ribosomes. As in other cells secreting glycoproteins, the Golgi (G') is large. Secretory granules (Gr') lie near the Golgi and accumulate in the peripheral cytoplasm. The basophilic granules, which are smaller than those of the acidophils, have been isolated and shown to contain gonadotrophic activity. The processes of hormone production and secretion are believed to be in general similar to those of the somatotroph.

From the pituitary of the rat
Magnification × 16,000

78

BARNES, B. G. Electron microscope studies on the secretory cytology of the mouse anterior pituitary. *Endocrinology, 71:*618 (1962).

FARQUHAR, M. G. Processing of secretory product by cells of the anterior pituitary gland. *In* Subcellular Organization and Function in Endocrine Tissues. I. H. Heller and K. Lederlis, editors. London, Cambridge University Press (1971) p. 79.

————, and RINEHART, J. F. Cytologic alterations in the anterior pituitary gland following thyroidectomy: an electron microscope study. *Endocrinology, 55:*857 (1954).

————. Electron microscopic studies of the anterior pituitary gland of castrate rats. *Endocrinology, 54:*516 (1954).

GOLUBOFF, L. G., MACRAE, M. E., EZRIN, C., and SELLERS, E. A. Autoradiography of tritiated thymidine labeled anterior pituitary cells in propylthiouracil treated rats. *Endocrinology, 87:*1113 (1970).

HARTLEY, M. W., MCSHAN, W. H., and RIS, H. Isolation of cytoplasmic pituitary granules with gonadotropic activity. *J. Biophysic. and Biochem. Cytol., 7:*209 (1960).

HYMER, W. C., and MCSHAN, W. H. Isolation of rat pituitary granules and the study of their biochemical properties and hormone activities. *J. Cell Biol., 17:*67 (1963).

KUROSUMI, K. Functional classification of cell types of the anterior pituitary gland accomplished by electron microscopy. *Arch. Histol. Jap., 29:*329 (1968).

LEZNOFF, A., FISHMAN, J., GOODFRIEND, L., McGARRY, E., BECK, J., and ROSE, B. Localization of fluorescent antibodies to human growth hormone in human anterior pituitary glands. *Proc. Soc. Exp. Biol. Med., 104:*232 (1960).

NAKANE, P. K. Classifications of anterior pituitary cell types with immunoenzyme histochemistry. *J. Histochem. Cytochem., 18:*9 (1970).

PELLETIER, G., and PUVIANI, R. Detection of glycoproteins and autoradiographic localization of [³H] fucose in the thyroidectomy cells of rat anterior pituitary gland. *J. Cell Biol., 56:*600 (1973).

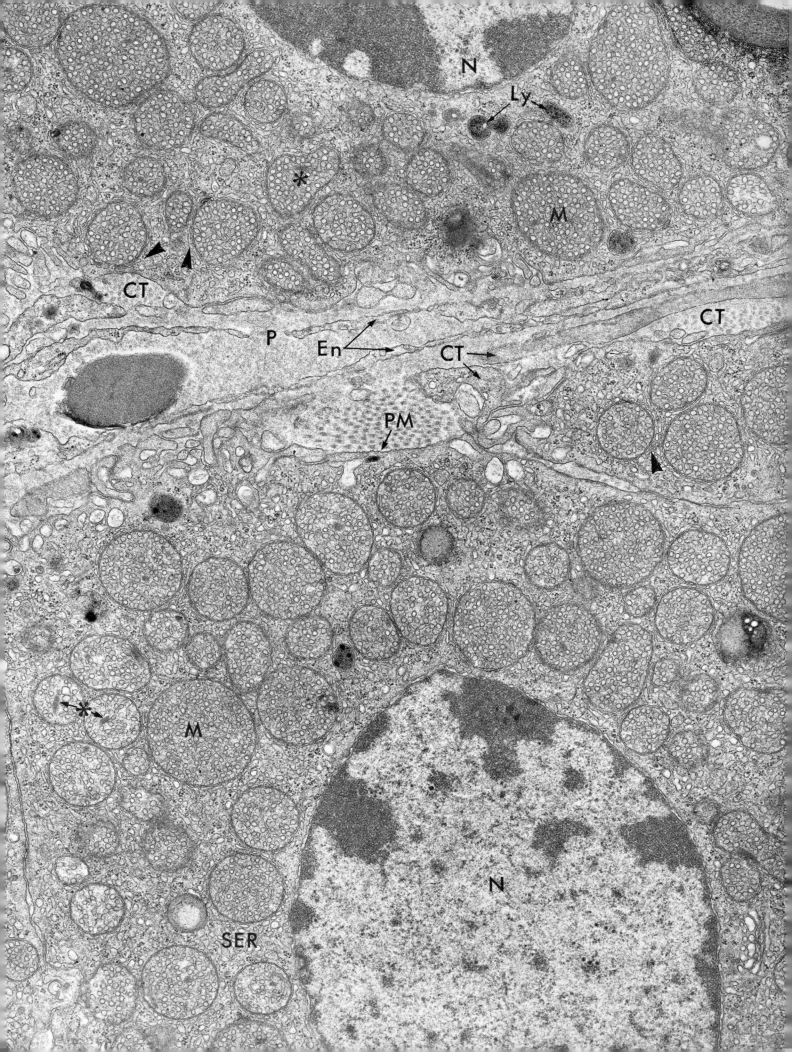

PLATE 18

The Zona Fasciculata of the Adrenal Cortex

The cells of the zona fasciculata of the adrenal cortex are responsible for the synthesis and secretion of a steroid hormone that plays a role in the regulation of protein and carbohydrate metabolism and exerts an inhibitory influence on the response of the connective tissues to inflammatory agents. In most species the hormone is cortisol, but in the rat fasciculata, shown here, corticosterone is produced. The deeper or innermost cortical region, the zona reticularis, apparently functions in a similar manner. However, the outermost layer of the gland, the zona glomerulosa, secretes aldosterone, which promotes sodium retention by the renal tubules.

Surgical separation and removal of each of the adrenal cortical zones is not feasible. Therefore other experiments had to be devised to discover the functional differences of zones that had long been seen under the light microscope. For instance, it has been found that when slices of adrenal cortex are incubated in appropriate medium *in vitro,* all the cortical hormones are secreted, and the amounts of each may be measured. However, if the connective tissue capsule together with the cells of the zona glomerulosa immediately beneath the capsule is removed, aldosterone production is markedly reduced, yet cortisol (corticosterone) production continues. Therefore the zona glomerulosa is apparently responsible for the synthesis of aldosterone, and the deeper lying fasciculata, for cortisol. Similar conclusions have been reached by observing the effects of hypophysectomy on the adrenal cortex. Without stimulation by ACTH, the fasciculata and reticularis atrophy, and the animal exhibits symptoms of cortisol (corticosterone) deficiency. The zona glomerulosa, on the other hand, is little affected by the removal of the pituitary, and aldosterone secretion continues.

Fine structural studies have revealed further distinguishing features of these cells (nucleus, N). Most striking is the "vesicular" structure of the cristae in the many mitochondria (M) in the zona fasciculata of the rat. It appears that tiny evaginations bud and separate from the inner limiting membrane so that the mitochondria contain numerous small vesicles in a matrix of moderate density. A small crystalline structure (·) is also frequently present. In other species, however, "vesicular" mitochondria have not been observed.

Rather, the mitochondrial cristae in most specimens have the tubular form characteristic generally of steroid-secreting cells (see Plate 19).

The endoplasmic reticulum in these cells is constructed entirely of anastomosing smooth-surfaced tubular elements (SER). Ribosomes, where present, lie free in the cytoplasmic ground substance. Close to the mitochondria, elements of the endoplasmic reticulum form a sheath around this organelle (arrowheads). A close physical relationship has also been observed between the mitochondria and lipid droplets.

Histophysiological studies of steroid-producing tissues are not nearly so complete as is generally the case for protein-secreting cells, but it is clear that the mitochondria, together with the endoplasmic reticulum, play a central role in hormone synthesis. The adrenal cortex of rats, unlike the interstitial cells of the testis (see Plate 19), produces very little cholesterol, the common precursor of steroid hormones. Rather, the required cholesterol is largely absorbed from the circulating fluids. The initial step in hormone synthesis—cleavage of the side chain of cholesterol to produce pregnenolone—occurs, as in other steroid-producing cells, inside the mitochondria. Subsequent steps, in which pregnenolone is modified to produce precursors of the cortical hormones, take place outside the mitochondria, mediated by certain enzymes associated with the endoplasmic reticulum. The final steps, in which the specific cortical hormones are produced, once again involve mitochondrial enzymes.

With the knowledge of the morphological relationship between the mitochondria and the cytoplasmic membranes, it is less surprising to find that hormone synthesis requires transfer of precursors between two cytoplasmic systems. The required transfers would seem to be facilitated by the close association of the membranes involved, which membranes apparently form a functional complex.

Electron micrographs have been of little help in understanding how the hormones, once formed, are secreted. The lipid droplets characteristic of the adrenal cortex contain cholesterol, primarily, and are therefore believed to store this important hormone precursor. Lysosomes (Ly) and accumulations of "wear and tear" pigments occur particularly in the zona reticularis, but neither they nor the lipid droplets are secretory granules. Instead it is believed that the steroid molecules diffuse to the periphery of the cell and then cross the irregular cell surface (PM) bordering the capillary endothelium (En). Once outside the cell of origin, the steroid molecules would find passage from the delicate layer of connective tissue (CT) into the circulation via fenestrae or pores (P) in the capillary endothelium.

This material was kindly provided by Dr. Robert Cardell.

From the adrenal of the rat
Magnification × 22,300

References

DEANE, H. W., and GREEP, R. O. A morphological and histochemical study of the rat's adrenal cortex after hypophysectomy, with comments on the liver. *Amer. J. Anat., 79*:117 (1946).

FAWCETT, D. W., LONG, J. A., and JONES, A. L. The ultrastructure of endocrine glands. *Recent Progr. Horm. Res., 25*:315 (1969).

FRIEND, D. S., and BRASSIL, G. E. Osmium staining of endoplasmic reticulum and mitochondria in the rat adrenal cortex. *J. Cell Biol., 46*:252 (1970).

————, and GILULA, N. B. A distinctive cell contact in the rat adrenal cortex. *J. Cell Biol., 53*:148 (1972).

GIACOMELLI, F., WIENER, J., and SPIRO, D. Cytological alterations related to stimulation of the zona glomerulosa of the adrenal gland. *J. Cell Biol., 26*:499 (1965).

GIROUD, C. J. P., STACHENKO, J., and VENNING, E. H. Secretion of aldosterone by the zona glomerulosa of rat adrenal glands incubated *in vitro*. *Proc. Soc. Exp. Biol. Med., 92*:154 (1956).

HUNT, T. E., and HUNT, E. A. The proliferative activity of the adrenal cortex using radioautographic technic with thymidine-H³. *Anat. Rec., 149*:387 (1964).

LONG, J. A., and JONES, A. L. The fine structure of the zona glomerulosa and the zona fasciculata of the adrenal cortex of the opossum. *Amer. J. Anat., 120*:463 (1967).

————, and JONES, A. L. Observations on the fine structure of the adrenal cortex of man. *Lab. Invest., 17*:355 (1967).

LUSE, S. Fine structure of adrenal cortex. *In* The Adrenal Cortex, A. B. Eisenstein, editor. Boston, Little Brown and Co. (1967) p. 1.

MACKAY, A. M. Atlas of human adrenal cortex ultrastructure. *In* Functional Pathology of the Human Adrenal Gland. T. Symington, editor. Edinburgh, E. and S. Livingstone, Ltd. (1969) p. 345.

RHODIN, J. A. G. The ultrastructure of the adrenal cortex of the rat under normal and experimental conditions. *J. Ultrastruct. Res., 34*:23 (1971).

SABATINI, D. D., and DEROBERTIS, E. D. P. Ultrastructural zonation of adrenocortex in the rat. *J. Biophysic. and Biochem. Cytol., 9*:105 (1961).

SHELTON, J. H., and JONES, A. L. The fine structure of the mouse adrenal cortex and the ultrastructural changes in the zona glomerulosa with low and high sodium diets. *Anat. Rec., 170*:147 (1971).

STACHENKO, J., and GIROUD, C. J. P. Functional zonation of the adrenal cortex: pathways of corticosteroid biogenesis. *Endocrinology, 64*:730 (1958).

PLATE 19

Interstitial Cells

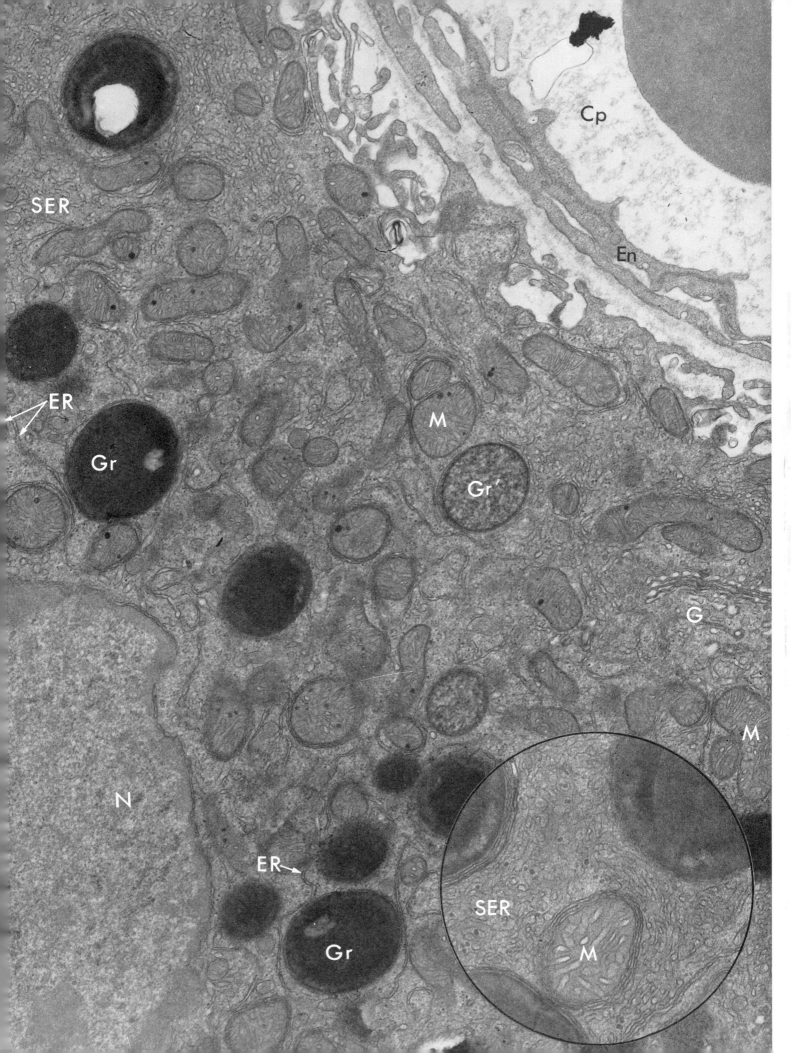

PLATE 19

Interstitial Cells

The testes have been known since ancient times to govern the development of secondary sexual characteristics of the male, an influence now known to be due to the production of the steroid, testosterone. Evidence available indicates that this hormone is secreted by the so-called interstitial cells, which are scattered among the connective tissue elements between the seminiferous tubules (see Plate 22) and which empty their secretion into nearby capillaries. This function of interstitial cells has been inferred from the observation that in certain individuals who produce no sperm both interstitial cells and secondary sexual characteristics remain normal.

A part of one interstitial cell is shown in this micrograph. At the lower left the nucleus (N) is surrounded by cytoplasm containing a Golgi region (G), mitochondria (M), and a highly developed endoplasmic reticulum. The latter two structures each display a particular morphology in steroid-secreting cells. As in the present example, examination of mitochondria frequently reveals their cristae to have a tubular rather than the more familiar shelflike form. The tubular cristae appear during functional differentiation of the tissue, and their number shows a positive correlation with the activity of the gland. Although the profiles of the ER in the perinuclear cytoplasm (ER) are long and slender and have associated ribosomes, the dominant elements of the ER in the more peripheral cytoplasm are without granules (SER) and seem to form a complex latticework of tubules (see inset, SER). This latter form of the ER is often extensively developed in steroid-secreting cells. Frequently, too, the smooth-surfaced cytoplasmic membranes are closely associated with mitochondria (see inset).

Cell fractionation studies have implicated the smooth-surfaced, tubular form of the endoplasmic reticulum in the elaboration of testosterone. In-

terstitial cells may possess SER in great abundance, and the membranes of this system are the chief component of the microsome fraction isolated from these cells. It is in this fraction that some of the enzymes taking part in the synthesis of cholesterol, a precursor of testosterone, are located. The testis is known to synthesize most of the cholesterol it requires (unlike adrenal cortical cells of the rat, Plate 18), so that there is a good correlation between biochemical and morphological observations on this point. As in the synthesis of the adrenal cortical hormones, the mitochondria are the site of conversion of cholesterol into pregnenolone. This steroid is then converted into testosterone in a series of enzymic reactions occurring in the microsome fraction. The participation of both SER and, as already mentioned, mitochondria is therefore essential, and the close association of these two membranous structures is a special feature of interstitial cells (see inset).

Large spherical granules (Gr), which stain strongly with osmium and lead, are also prominent components of the cytoplasm. Another type of large granule (Gr′), also displayed here, is probably a variant form of the denser ones. It is thought that the granules all contain cholesterol and other lipids that represent stored hormonal precursors. In contrast, the final secretory product, the hormone, is not stored within such granules or indeed in any other kind of granule prior to secretion. Instead the hormone probably diffuses from the cell in molecular form. Once in the intercellular spaces, the testosterone must penetrate the basement membrane and thin endothelium (En) of neighboring capillaries (Cp) to be carried to target organs in other parts of the body.

From the testis of the mouse
Magnification × 29,000
Inset × 55,000

85

BAILLIE, A. H. Further observations on the growth and histochemistry of the Leydig tissue in the postnatal prepubertal mouse testis. *J. Anat., 98:*403 (1964).

BELT, W. D., and CAVAZOS, L. F. Fine structure of the interstitial cells of Leydig in the boar. *Anat. Rec., 158:*333 (1967).

CHRISTENSEN, A. K. The fine structure of testicular interstitial cells in guinea pigs. *J. Cell Biol., 26:*911 (1965).

———, and FAWCETT, D. W. The fine structure of testicular interstitial cells in mice. *Amer. J. Anat., 118:*551 (1966).

———, and GILLIM, S. W. The correlation of fine structure and function in steroid-secreting cells, with emphasis on those of the gonads. *In* The Gonads. K. W. McKerns, editor. New York, Appleton-Century-Crofts (1969) p. 415.

DE KRETSER, D. M. The fine structure of the interstitial cells in men of normal androgenic status. *Z. Zellforsch., 80:*594 (1967).

———. Changes in the fine structure of the human testicular interstitial cells after treatment with human gonadotrophins. *Z. Zellforsch., 83:*344 (1967).

FAWCETT, D. W., and BURGOS, M. H. Studies on the fine structure of the mammalian testis. II. The human interstitial tissue. *Amer. J. Anat., 107:*245 (1960).

FRANK, A. L., and CHRISTENSEN, A. K. Localization of acid phosphatase in lipofuscin granules and possible autophagic vacuoles in interstitial cells of guinea pig testis. *J. Cell Biol., 36:*1 (1968).

MUROTA, S., SHIKITA, M., and TAMAOKI, B. Intracellular distribution of the enzymes related to androgen formation in mouse testes. *Steroids, 5:*409 (1965).

PLATE 20

The Follicular Epithelium of the Thyroid

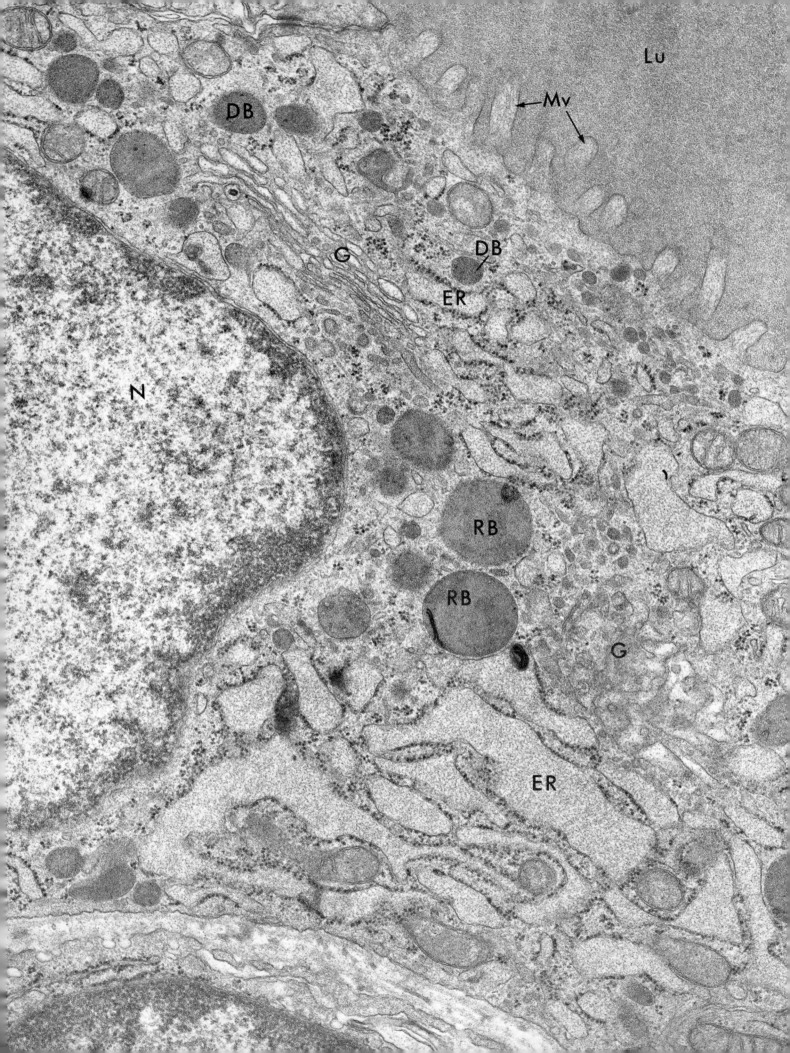

PLATE 20

The Follicular Epithelium of the Thyroid

The secretory epithelium of the thyroid, a derivative of the vertebrate pharynx, has a profound effect on development in many of the vertebrate phyla. In amphibians, for example, it plays an essential role in the metamorphosis from the larval to the adult form. If untreated, hypothyroidism in young humans results in irreparable damage to the nervous system and radical impairment of growth (cretanism). In adults the thyroid hormones regulate the rate of metabolism, and their proper functioning is necessary for normal life.

The follicle is the basic unit of structure in thyroid tissue. A single sheet of epithelial cells encloses a spherical cavity filled with the glycoprotein, thyroglobulin, which serves to store the thyroid hormones in inactive form. Upon appropriate stimulation, the epithelial cells, such as the one shown in part in this micrograph, bring about the release of the two thyroid hormones from the thyroglobulin or colloid. These then pass from the follicular lumen (Lu), across the plasma membrane limiting the microvilli (Mv) that cover the luminal cell surface, and through the epithelium to gain access to underlying vascular elements. In the production of the colloid, the thyroid may be considered an exocrine gland, but in the release of hormone into the circulatory system it functions as an endocrine gland.

The thyroglobulin is quite rich in leucine, and by injecting this amino acid in radioactive form the site of protein synthesis as well as the cytological events in secretion may be followed. This is technically achieved by preparing autoradiographs at increasing time intervals after exposure to the label. Information from such experiments indicates that protein production in these cells has much in common with that in pancreatic acinar cells (see text figure 11a) and other protein-secreting cells. Ten minutes after administration the label is detectable in association with the rough-surfaced endoplasmic reticulum (ER). The ribosomes are, of course, believed to be the sites of synthesis, and the protein is then thought to be sequestered within the membrane-bounded cisternae. It can be observed here that the ER elements are extensive in normal thyroid epithelium and seem frequently to be distended with flocculent material, presumably protein. After an hour, Golgi (G) elements (seen here nearby the nucleus, N), become labeled. Inasmuch as thyroglobulin is a glycoprotein it seems likely that the polysaccharide moiety is manufactured or united with the protein by enzymes within these membranes (cf. Plates 12 and 23). Later the label moves into the apical cytoplasm and finally into the lumen. Secretion droplets comparable to those present in pancreatic exocrine cells (Plate 11) or in goblet cells (Plate 12) have not been identified beyond question. The recognition of secretory mechanisms is made the more difficult by the presence of numerous colloid reabsorption droplets (see below) in the apical cytoplasm as well as elsewhere in the cell. There remains, therefore, some doubt as to how the globulin leaves the follicle cell.

The thyroid hormone, thyroxine, is formed in the follicle cells through the iodination of the amino acid, tyrosine. Circulating iodide ion is taken up selectively by the cells and oxidized, whereupon the released iodine combines with tyrosine to produce diiodotyrosine. Molecules of this compound undergo an oxidative coupling to form triiodothyronine or tetraiodothyronine, which exist finally in combination with globulin in the follicle. It can be shown with radioactive iodide that iodide in one form or another accumulates within the follicular epithelium and is thereafter linked to thyroglobulin, the colloid, at or near the apical cell surface.

In hypophysectomized animals colloid distends the follicles, and circulating thyroid hormones drastically decrease. The gland is then at a low level of endocrine activity. By restimulation of the gland by means of TSH (thyroid-stimulating hormone from the anterior pituitary) injected intravenously, the mechanism of hormone release may be studied. The apical cell surface expands, developing blebs on which are fine pseudopodia-like extensions. These seem to engulf portions of the material within the lumen, leading to the formation of intracellular "colloid droplets." This is in effect phagocytosis of the colloid and is followed by migration of the membrane-bounded droplets into deeper areas of the cytoplasm. There they are associated with small dense bodies (DB) that have been shown to contain acid phosphatase, one of the enzymes usually found in lysosomes. As time goes on, the colloid droplets also come to exhibit acid phosphatase activity, resulting, it is thought, from union of the small dense bodies with the larger colloid droplets. Thus the colloid droplets, which are known to contain material that is PAS positive (as is the colloid) become large dense bodies in which lysosomal enzymes are apparently degrading the thyroglobulin and thus bringing about the release of hormones. These small molecules eventually find their way into the underlying capillaries, while dense granular structures, "residual bodies," remain as a sort of debris after a cycle of "reabsorption" of thyroglobulin has run its course. Some of the larger dense granular structures seen in this plate (RB) may represent such evidences of previous hormone release. A lysosomal fraction isolated from thyroid epithelium has been demonstrated to be particularly effective in lysing the thyroglobulin. This evidence has therefore substantiated the role of lysosomes in hormone release.

From the thyroid of the rat
Magnification × 37,000

References

BOUCHILLOUX, S., CHABAUD, O., MICHEL-BÉCKET, M., FERRAND, M., and ATHORIËL-HAON, A. M. Differential localization in thyroid microsomal subfractions of a mannosyltransferase, two N-acetyl-glycosaminyltransferases and a galactosyltransferase. *Biochem. Biophys. Res. Commun., 40*:314 (1970).

BRADLEY, A. S., and WISSIG, S. L. The anatomy of secretion in the follicular cell of the thyroid gland. III. The acute effect *in vivo* of thyrotropic hormone on amino acid uptake and incorporation into protein by the mouse thyroid gland. *J. Cell Biol., 30:*433 (1966).

EKHOLM, R., and SJÖSTRAND, F. S. The ultrastructural organization of the mouse thyroid gland. *J. Ultrastruct. Res., 1:*178 (1957).

———, and SMEDS, S. On dense bodies and droplets in the follicular cells of the guinea pig thyroid. *J. Ultrastruct. Res., 16:*71 (1966).

HADDAD, A., SMITH, M. D., HERSCOVICS, A., NADLER, N. J., and LEBLOND, C. P. Radioautographic study of *in vivo* and *in vitro* incorporation of fucose-³H into thyroglobulin by rat thyroid follicular cells. *J. Cell Biol., 49*:856 (1971).

HERSCOVICS, A. Biosynthesis of thyroglobulin. Incorporation of (1-¹⁴C) galactose, (1-¹⁴C) mannose and (4, 5-³H₂) leucine into soluble proteins by rat thyroids *in vitro. Biochem. J., 112*:709 (1969).

———. Biosynthesis of thyroglobulin: Incorporation of (³H) fucose into proteins by rat thyroids *in vitro. Biochem. J., 117*:411 (1970).

KLINCK, G. H., OERTEL, J. E., and WINSHIP, I. Ultrastructure of normal human thyroid. *Lab. Invest., 22*:2 (1970).

NADLER, N. J., YOUNG, B. A., LEBLOND, C. P., and MITMAKER, B. Elaboration of thyroglobulin in the thyroid follicle. *Endocrinology, 74:*333 (1964).

SELJELID, R. On the origin of colloid droplets in thyroid follicle cells. *Exp. Cell Res., 41:*688 (1966).

SHELDON, H., McKENZIE, J. M., and VAN NIMWEGAN, D. Electron microscopic autoradiography. The localization of I¹²⁵ in suppressed and thyrotropin-stimulated mouse thyroid gland. *J. Cell Biol., 23:*200 (1964).

WETZEL, B. K., SPICER, S. S., and WOLLMAN, S. H. Changes in fine structure and acid phosphatase localization in rat thyroid cells following thyrotropin administration. *J. Cell Biol., 25* (no. 3, part 1)*:*593 (1965).

WHUR, P., HERSCOVICS, A., and LEBLOND, C. P. Radioautographic visualization of the incorporation of galactose-³H and mannose-³H by rat thyroids *in vitro* in relation to the stages of thyroglobulin synthesis. *J. Cell Biol., 43:*289 (1969).

PLATE 21
The Ovarian Follicle

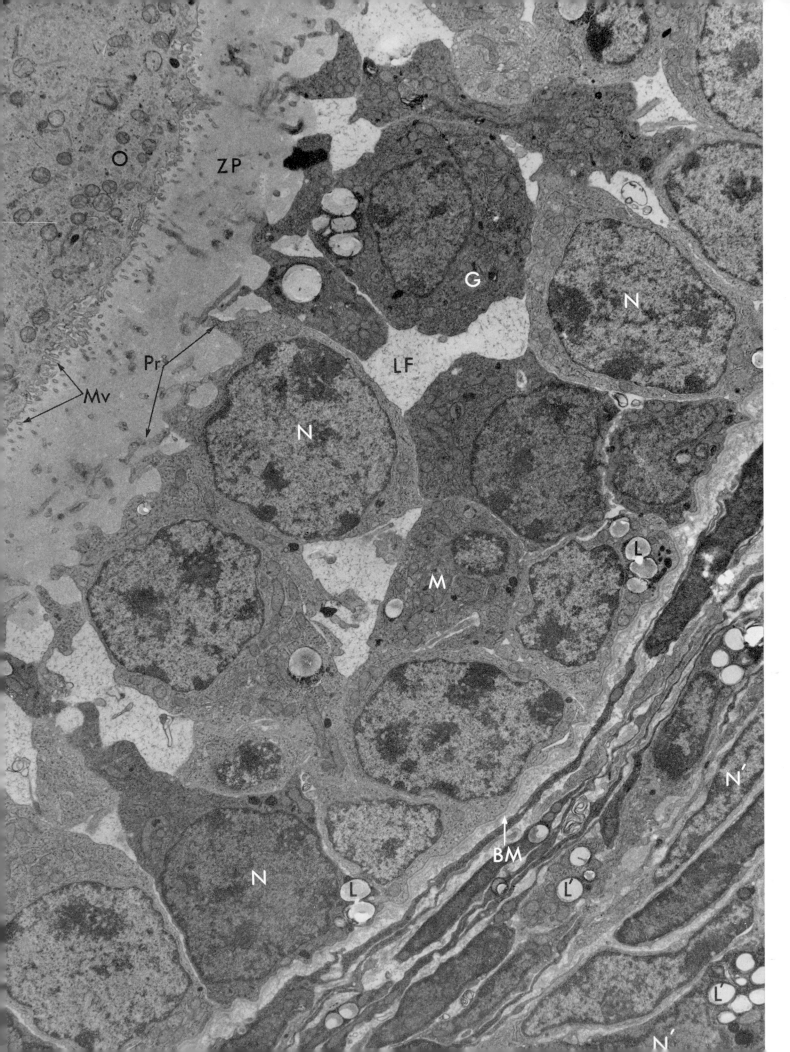

PLATE 21

The Ovarian Follicle

The cells of the ovarian follicle, shown in this low power electron micrograph, line a cystlike space in which the oocyte (O) or egg matures. The follicle cells (nuclei, N) originate by the budding off of a group of cells from a cortical ovarian layer, the germinal epithelium. One of the cells in the group becomes the ovum and together with the surrounding cells constitutes the primary follicle. Such follicles are responsive to hormonal control, but in sexually mature mammals only certain primary follicles respond to follicle-stimulating hormone, FSH, produced by the anterior pituitary (see Plate 17). Oocyte and follicular epithelium then undergo an extended period of maturation. As part of this, the stromal tissue of the ovary is organized around the spherical epithelial mass to form the theca. The entire structure, follicle and theca, produces not only the mature egg cell but also two hormones essential to the maintenance of the normal reproductive cycle.

In the upper left-hand portion of this micrograph, a small part of the egg cell (O) is seen. During development it acquires an extensive cytoplasm and in nonplacental animals a large amount of yolk. While the cytoplasm of male germ cells is specialized for locomotion (see Plate 22), the egg cytoplasm becomes a storehouse of materials needed for the earliest developmental stages after fertilization. In vertebrates there is evidence that yolk protein is manufactured for the most part in the liver, transported through the circulatory system to the egg, and then incorporated into the egg cell without detectable change. This last step is probably accomplished by a process of pinocytosis; that is, the protein reaches the surface of the egg where it enters small pitlike structures or indentations of the plasma membrane. The pits bud off, thus forming intracytoplasmic vesicles containing the yolk protein. Some yolk formation has also been detected within the egg itself.

Although the follicular epithelium encloses the egg, there is no indication that it contributes directly to the growth or nourishment of the germ cell. As development proceeds, it becomes separated from the egg by a layer of homogeneous material, the zona pellucida (ZP). This carbohydrate-rich substance is probably produced by both the egg and the follicular cells. It is penetrated by the microvilli (Mv) of the egg and by processes (Pr) extending from the epithelium. Desmosome-like structures have been observed where the projections from the oocyte and follicular cells are in contact. The follicular cells are also separated from each other by the fluid liquor folliculi (LF), which apparently originates from them.

The cytoarchitecture of the follicular epithelial cells suggests in fact that they are synthetically active. Abundant mitochondria (M), Golgi regions (G), and dense concentrations of ribosomes, both attached and unattached to ER cisternae, are present. Lipid droplets (L) are a constant feature of these cells and may serve as storage depots for materials involved in steroid synthesis (see below).

The follicular epithelium rests on a basement membrane (BM) and as already indicated is surrounded by the theca. The cells of the theca interna (with nuclei N'), that is, those lying nearest the follicle, resemble fibrocytes in certain respects. They are flattened cells forming membranous layers and are separated by fibrous intercellular material. They have, however, two distinguishing features: they contain lipid droplets (L'), and their mitochondria resemble those found in steroid-secreting cells, such as those of the interstitial tissue (Plate 19) and the adrenal cortex (Plate 18). The theca cells are probably the site of production of estrogenic hormone, which is known to be manufactured by the ovary. While the epithelial cells may also produce estrogen, the evidence for this is less convincing. The theca and possibly the growing follicle act therefore as endocrine organs. During the early stages of the ovulatory cycle the thecal cells secrete estrogen. When this hormone builds up to sufficiently high levels in the circulation, the anterior pituitary secretes LH, luteinizing hormone. This in turn brings about ovulation and the further maturation of the follicle.

At ovulation the egg cell and immediately surrounding follicular cells (the corona radiata) are expelled from the ovary after the epithelial cells become loosened from their neighbors. The remaining follicular cells and the theca interna undergo further development, and both contribute to the formation of the corpus luteum. The latter exists for a limited time as an endocrine organ secreting progesterone, the steroid hormone that is essential in the preparation of the uterus for the implantation of the fertilized ovum.

From the ovary of the mouse
Magnification × 8,000

References

ADAMS, E. C., and HERTIG, A. T. Studies on guinea pig oocytes. I. Electron microscopic observations on the development of cytoplasmic organelles in oocytes of primordial and primary follicles. *J. Cell Biol., 21:*397 (1964).

ANDERSON, E. The formation of the primary envelope during oocyte differentiation in teleosts. *J. Cell Biol., 35:*193 (1967).

———, and BEAMS, H. W. Cytological observations on the fine structure of the guinea pig ovary with special reference to the oogonium, primary oocyte and associated follicle cells. *J. Ultrastruct. Res., 3:*432 (1960).

BARKER, W. L. A cytochemical study of lipids in sows' ovaries during the estrous cycle. *Endocrinology, 48:*772 (1951).

DEANE, H. W. Histochemical observations on the ovary and oviduct of the albino rat during the estrous cycle. *Amer. J. Anat., 91:*363 (1952).

HERTIG, A. T., and ADAMS, E. C. Studies on the human oocyte and its follicle. I. Ultrastructural and histochemical observations on the primordial follicle stage. *J. Cell Biol., 34:*647 (1967).

KNIGHT, P. F., and SCHECHTMAN, A. M. The passage of heterologous serum proteins from the circulation into the ovum of the fowl. *J. Exp. Zool., 127:*271 (1954).

ROTH, T. F., and PORTER, K. R. Yolk protein uptake in the oocyte of the mosquito *Aedes aegypti* L. *J. Cell Biol., 20:*313 (1964).

SOTELO, J. R., and PORTER, K. R. An electron microscope study of the rat ovum. *J. Biophysic. and Biochem. Cytol., 5:*327 (1959).

PLATE 22

The Germinal Epithelium of the Male

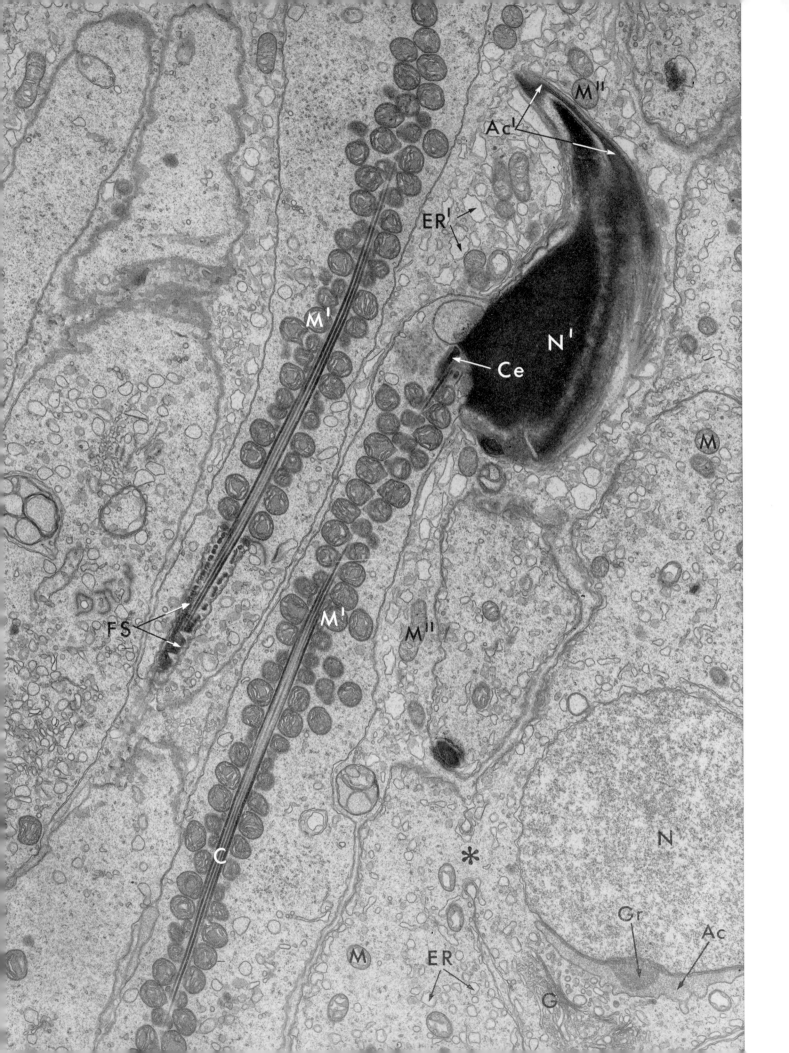

PLATE 22

The Germinal Epithelium of the Male

The male germ cells are produced in the testis by the epithelium of the seminiferous tubules. Undifferentiated spermatogonia, which form part of the basal cell layer of this epithelium, undergo repeated mitotic divisions. After a period of growth, each resulting spermatocyte enlarges and then divides meiotically to produce four spermatids. These in turn pass through a complex metamorphosis during which, by a process called spermiogenesis, they are transformed into mature sperm. It is, of course, impossible to document this intricate life history with one micrograph. However, the present illustration does show two stages in spermatid development and, by illustrating extremes, demonstrates how drastic this transformation is. At lower right one observes the profile of a spherical nucleus (N) that belongs to a fairly early spermatid. At one pole it is covered by a flattened vesicle, the developing acrosome (Ac), which contains a dense granule (Gr), the proacrosomal granule. The cap and the granule arise from the Golgi complex (G), and together they form the anterior tip of the mature sperm (see Plate 23). Early in spermiogenesis the cytoplasm is extensive. Within its irregular outline, mitochondria (M) and vesicles of the endoplasmic reticulum (ER) may be seen. At this point in development, the four spermatids remain linked by cytoplasmic bridges, one of which appears in this micrograph (*).

The appearance of a nearly mature sperm is illustrated at the center of the micrograph. The head, which is hook-shaped in the mouse, contains within its nucleus (N') the hereditary material in dense, compact form. The anterior tip of the nucleus is covered by the acrosome cap (Ac'), which lies just beneath the plasma membrane. The cap is believed to contain the enzyme hyaluronidase, which may aid the penetration of the egg by the sperm during fertilization.

No less remarkable in this metamorphosis are the changes in the spermatid cytoplasm, which essentially transforms into an elongate motile appendage with which the sperm can swim. Two centrioles are involved in the formation of the neck region, and of these one (Ce) initiates the development of the long bundle of filaments (C) that is the core of the middle piece and tail. This core comprises a 9 + 2 array of microtubules and so, in its internal structure, is very similar to the cilium (see Plate 8). In the middle piece the core of the flagellum is spirally wrapped by a sheath constructed of mitochondria. Oblique sections (M') of the spiral reveal the mitochondrial nature of this structure.

Posterior to the middle piece, in the tail proper, a fibrous sheath, a portion of which may be examined in this micrograph (FS), is wound around the axial filamenture. Only at its distal end does the flagellum lack a special covering. Although a sheath of cytoplasm covers the middle piece as it develops, this sheath is later lost, so that at maturity the plasma membrane is closely applied to the axial structures.

In the final stages of spermiogenesis, the spermatids become embedded in the Sertoli cells. Many germ cells, each with its anterior tip oriented toward the periphery of the seminiferous tubule, are thus anchored to the epithelium, while their tails lie free in the lumen. Around the dense sperm head and middle pieces shown in this plate there are portions of a Sertoli cell, which contain mitochondria (M") and dilated vesicles of the endoplasmic reticulum (ER'). No protoplasmic continuity between the Sertoli cell and the germ cells has been observed: each cell type is surrounded by an uninterrupted plasma membrane. Sertoli cells are often called nurse cells, but though they support the developing sperm, it is not really known what additional functions they perform.

From the testis of the mouse
Magnification $\times$ 13,500

BURGOS, M. H., and FAWCETT, D. W. Studies on the fine structure of the mammalian testis. I. Differentiation of the spermatids in the cat (*Felis domestica*). *J. Biophysic. and Biochem. Cytol., 1:*287 (1955).

DE KRETSTER, D. M. Ultrastructure features of human spermiogenesis. *Z. Zellforsch. Mikrosk. Anat., 98*: 477 (1969).

FAWCETT, D. W. Sperm tail structure in relation to the mechanism of movement. *In* Spermatozoan motility. D. W. Bishop, editor. Washington, D. C., AAAS Publication No. 72 (1962) p. 147.

————. The topographical relationship between the plane of the central pair of flagellar fibrils and the transverse axis of the head in guinea-pig spermatozoa. *J. Cell Sci., 3:*187 (1968).

————. A comparative view of sperm ultrastructure. *Biol. Reprod., 2* (Suppl.):90 (1970).

————, and ITO, S. The fine structure of bat spermatozoa. *Amer. J. Anat., 116:*567 (1965).

————, and PHILLIPS, D. M. The fine structure and development of the neck region of the mammalian spermatozoon. *Anat. Rec. 165:*153 (1969).

————, and PHILLIPS, D. M. Observations on the release of spermatozoa and on changes in the head during passage through the epididymis. *J. Reprod. Fert., 6* (Suppl.): 405 (1969).

HORSTMANN, E. Elektronmikroscopische Untersuchungen zur Spermiohistogenese beim Menschen. *Z. Zellforsch., 54:*68 (1961).

PHILLIPS, D. M. Insect sperm: their structure and morphogenesis. *J. Cell Biol., 44:*243 (1970).

————. Comparative analysis of mammalian sperm motility. *J. Cell Biol., 53:*561 (1972).

YASUZUMI, G., TANAKA, H., and TEZUKA, O. Spermatogenesis in animals as revealed by electron microscopy. VIII. Relation between the nutritive cells and the developing spermatids in a pond snail, *Cipangopaludina malleata* Reeve. *J. Biophysic. and Biochem. Cytol., 7:* 499 (1960).

ZAMBONI, L., and STEFANINI, M. The fine structure of the neck of mammalian spermatozoa. *Anat. Rec., 169:*155 (1971).

————, ZEMJANIS, R., and STEFANINI, M. The fine structure of monkey and human spermatozoa. *Anat. Rec., 169:*129 (1971).

PLATE 23

Formation of the Acrosome

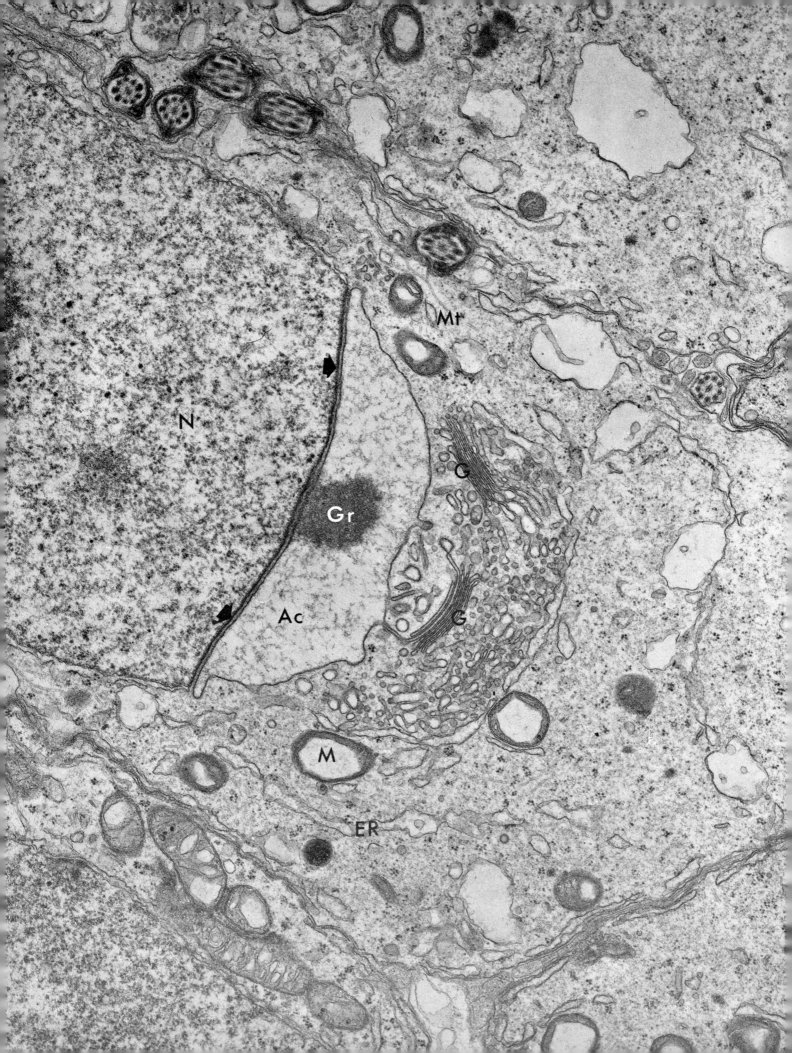

PLATE 23

Formation of the Acrosome

In almost all types of animals studied—both invertebrate and vertebrate—the anterior region of male germ cells bears specialized structures that play a role in penetration of the egg during fertilization. These are essential because egg cells are generally covered by membranous and gelatinous coatings and in mammals are surrounded by a layer of follicle cells (see Plate 21) that remain with the egg after its ovulation. To aid in overcoming these barriers is apparently the function of the acrosome, a dense structure that caps the sperm nucleus and is closely covered by the plasma membrane of the sperm cell. In the mature sperm head (see Plate 22), therefore, the acrosome is the chief cytoplasmic structure present, although thin layers of cytoplasmic ground substance may sometimes be detected between it and the nucleus.

The formation of the acrosome was carefully studied under the light microscope by R. H. Bowen, who learned (about forty-five years ago) that the acrosome originated in the Golgi region. The details of acrosome formation have been seen clearly with the electron microscope, and confirmation of the importance of the Golgi in this process has been obtained. Specific features of acrosome development may differ from species to species, but in general the events have the following common aspects, some of which may be seen in the mouse spermatid (nucleus, N) illustrated in this micrograph. First a proacrosomal granule (Gr) appears within one of the Golgi vesicles. Close association between Golgi cisternae (G) and the acrosomal vesicle (Ac) is maintained as the vesicle and its contents increase in size. The vesicle then becomes closely associated with the apical pole of the nucleus. The internal cavity of the nuclear envelope is reduced to a thin layer of constant width (arrows), and the envelope and acrosomal vesicle are separated by a thin cytoplasmic layer.

In the young spermatid the Golgi, together with the developing acrosome, is surrounded by abundant cytoplasm, within which elements of the endoplasmic reticulum (ER), ribosomes, and microtubules (Mt) may be identified. The mitochondria (M), which are also present, display a large internal cavity, an early change in a series that will lead to the formation of the mito-chondrial tail sheath of the mature sperm (see Plate 22). Once the Golgi has played its role, however, it separates from the acrosome and migrates to the posterior part of the cell. The cytoplasm also moves caudally, and finally the maturing acrosome lies just beneath the plasma membrane. The acrosomal vesicle then becomes filled with dense material and takes on a shape characteristic of the species.

When sperm and egg come into contact, a series of events termed the acrosome reaction occurs. This has been studied in detail in certain marine invertebrates where it is easy to obtain stages in the reaction. In animals in which internal fertilization occurs, observation of the process is difficult, but it seems fairly certain that its general features are similar. In brief, the plasma membrane and the apical membrane of the acrosome perforate, thereby releasing the acrosomal contents. The edges of the perforation fuse, so that the acrosomal membrane becomes continuous with the plasma membrane and thus part of the apical surface of the sperm. From this surface a projection forms, an eversion of what was the acrosomal membrane. In vertebrates the projection contains a rodlike structure, the perforatorium, which grows out within it. It is at this specialized tip that the sperm makes contact with the egg plasma membrane.

Naturally, it has been of interest to discover the chemical nature of the acrosome. The proacrosomal granule and the material within the fully developed acrosome stain positively with the periodic acid-Schiff (PAS) method, an indication of the presence of polysaccharide. Further histochemical investigations, as well as analyses of isolated acrosomes, have suggested that a glycolipid—or possibly a glycoprotein—is the chief PAS-positive component.

A number of methods are known for bringing about the acrosomal reaction in sperm suspension. When this reaction occurs, a marked increase in lytic activity, due to the presence of the enzyme hyaluronidase, may be detected in the suspending medium. It is believed that this enzyme enables the sperm to break down egg coatings, particularly the intercellular substance (hyaluronic acid) that binds together the follicle cells surrounding the mammalian egg (see Plate

21). The contents of the acrosome therefore constitute a special secretory product. Like other instances of elaboration of polysaccharide secretions (see Plate 12) the Golgi membranes play a role at least in the final assemblage and packaging of the product.

From the testis of the mouse
Magnification × 32,000

References

BOWEN, R. H. On the acrosome of the animal sperm. *Anat. Rec., 28:*1 (1924).

COLWIN, L. H., and COLWIN, A. L. Membrane fusion in relation to sperm-egg association. *In* Fertilization. Comparative Morphology, Biochemistry and Immunology. C. B. Metz and A. Monroy, editors. New York, Academic Press, vol. I (1967) p. 295.

DAN, J. C. Acrosome reaction and lysins. *In* Fertilization. Comparative Morphology, Biochemistry and Immunology. C. B. Metz and A. Monroy, editors. New York, Academic Press, vol. I (1967) p. 237.

————, and HAGIWARA, Y. Studies on the acrosome. IX. Course of acrosome reaction in the starfish. *J. Ultrastruct. Res., 18:*562 (1967).

HARTREE, E. F., and SRIVASTAVA, P. N. Chemical composition of the acrosomes of ram spermatozoa. *J. Reprod. Fertil., 9:*47 (1965).

KAYE, J. S. Acrosome formation in the house cricket. *J. Cell Biol., 12:*411 (1962).

PLATE 24
Cells of the Proximal Convoluted Tubule

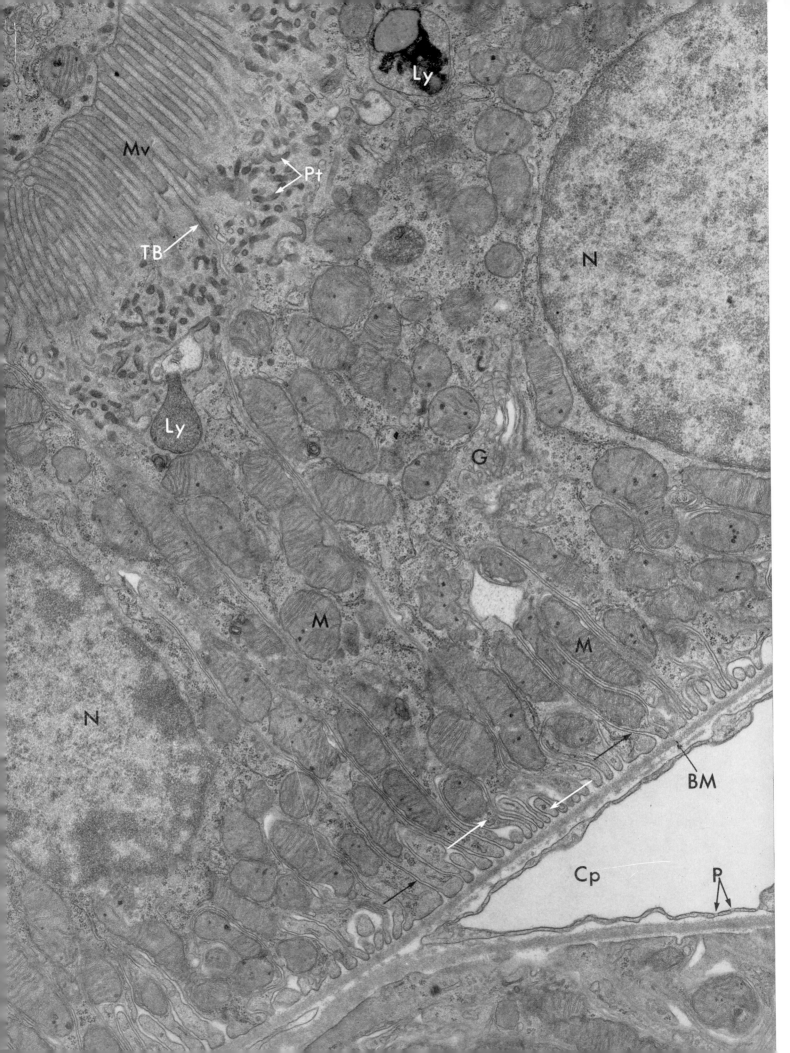

PLATE 24
Cells of the Proximal Convoluted Tubule

AFTER its formation by the renal corpuscle (Plate 25), the glomerular filtrate passes into the lumen of the proximal convoluted tubule. It is at this site that reabsorption of a number of essential metabolites and of substantial amounts of water occurs for the maintenance of homeostasis. Transport of these materials from the lumen of the nephron across the tubular epithelium may be active—that is, requiring energy to overcome unfavorable concentration gradients —or it may be passive, involving diffusion along a gradient. Glucose and sodium ion are two important metabolites that are actively reabsorbed by the proximal convoluted tubule, so that normally all the glucose and most of the sodium are conserved. As a result of the gradients established by these actively transported particles, chloride ion and water also move through the cells of the tubule. In addition, the tubules passively reabsorb some urea, the principal nitrogenous waste product of mammals, but their relative impermeability to this organic molecule seems to prevent its retention in appreciable amounts.

The simple cuboidal or low columnar epithelium lining the proximal convoluted tubule is, as indicated above, highly active, and its structure reflects this fact. As one would expect, many mitochondria (M) are found in the hard-working cells of the proximal tubule; their profiles are observed lateral to the nuclei (N), near the Golgi complex (G), and close to the membranes in the basal zone of the cells, where energy is required for active transport.

In addition, both the apical and basal surfaces of each cell, the areas where exchanges occur between the cell and its environment, are expanded in surface area. As in the case of the intestinal absorptive epithelium (see Plates 6 and 7), the apical surface displays numerous, closely packed microvilli (Mv). In life they project into the open lumen of the nephron, which, as here, is often occluded when the nephron collapses during EM preparative procedures. The basal surface of the plasma membrane, on the other hand, is notable for its deep infoldings (black arrows) that limit narrow columns of cytoplasm within which the mitochondria are enclosed. It is thought that cytoplasmic processes from adjacent tubule cells interdigitate at the basal surface much as do the podocytes of the glomerular epithelium (see Plate 25). Thus in thin sections the profiles of isolated protoplasmic feet (white arrows, at lower right) are customarily observed bordering the subjacent basal lamina (BM).

It is significant that transport is through the cells of the tubule rather than between them. The terminal bars (TB) and associated "tight junctions" apparently seal off the intercellular spaces from the lumen of the tubule.

At the surfaces of the microvilli the essential steps in glucose transport are believed to occur. The theory most favored currently proposes the existence of a special type of carrier molecule that effects transfer of the sugar across the membrane into the cytoplasm. There the carrier is separated from the glucose, which is trapped within the cell, while the carrier is released for further transport activity. Sugar molecules accumulated within the cell must then cross the folded basal cell surface (black arrows), the underlying basement membrane (BM), and the thin endothelium of the nearby capillary (Cp) in order to reenter the circulatory system. It may be noted that this endothelium is characterized by the presence of pores (P) or fenestrae, which are, however, not open. Instead, a single-layered membrane or diaphragm is present in them. The physiological significance of these pores is unknown.

In the case of sodium reabsorption, however, it is the basal cell surface that is believed to be the site where Na^+ is actively transported out of the cell, probably in exchange for K^+. This hypothetical "sodium-potassium pump" mechanism, originally proposed to explain ionic movements in nerve cells, is assumed to exist here. Thus as Na^+ is pumped out of the cell via the basal surface, diffusion of Na^+ from the tubule lumen into the cells is encouraged by a favorable concentration gradient.

Occasionally large molecules are present within the lumen of the tubule, as, for example, proteins that have leaked through a diseased glomerulus or dense marker substances, such as hemoglobin or ferritin, that have been introduced experimentally. The cells dispose of these abnormal particles much as do the cells of the intestinal epithelium (see Plate 7). Electron microscopists have observed that such materials are funneled between the microvilli and into deep slender pits or wells. These terminate in tiny pouches, which seem especially developed for the uptake of large molecules from the filtrate. Profiles of these tubular invaginations (Pt) are abundant in the apical cytoplasm. When filled, their terminal pouches apparently pinch off, and the vesicles thus formed migrate into the depths of the cell cortex. There they sometimes fuse to form larger vesicles, and their contents blend into larger granules. Recent histochemical studies show that these membrane-bounded bodies undergo further transformation, taking on the properties of lysosomes (Ly). It is therefore assumed that their contents are progressively hydrolyzed.

From the kidney of the mouse
Magnification × 21,000

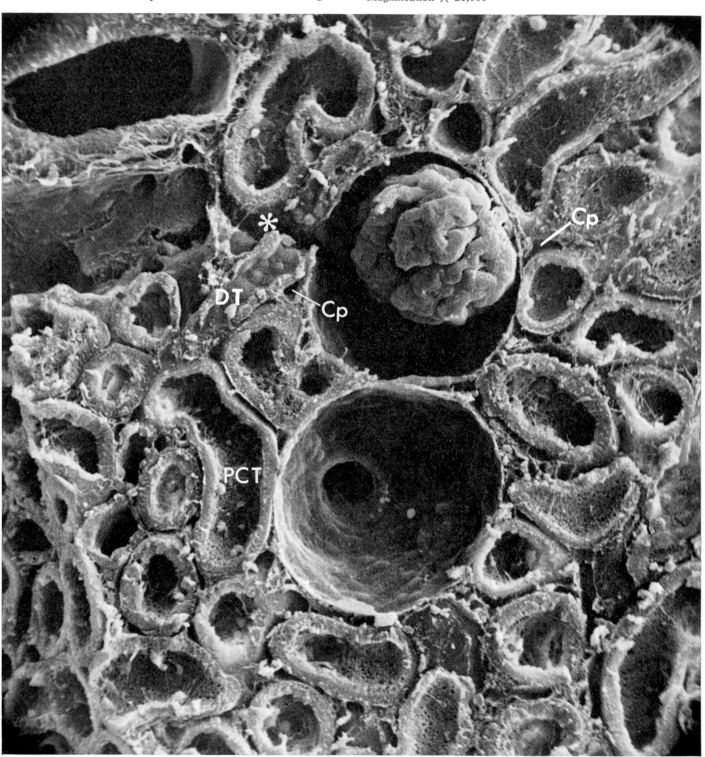

Text Figure 24a

Scanning Image of the Kidney Cortex

Scanning Image of the Kidney Cortex
(Text Figure 24a)

This micrograph, taken with the scanning electron microscope, shows the structure of the kidney cortex. The surface depicted was exposed by cutting the kidney into two or more pieces after it had been fixed by perfusion with glutaraldehyde.

Certain structural features of the cortex are immediately recognizable. The hemispherical cavity with a "hole" in the bottom represents a Bowman's capsule with the glomerulus removed. The observer is looking at the free surface of the thin parietal epithelium. The "hole" is the opening into a proximal tubule, which is of the same nature as those present in large numbers in the surrounding cortex. Just adjacent (and above, in the micrograph) to the open capsule is another corpuscle with the glomerulus still in place. The urinary space around the glomerulus seems large and may have been rendered so by fixation. Viewed from its visceral surface the glomerular tuft appears as an irregular convolution of worms. These are the capillaries depicted in Plate 25. Here, of course, they are covered by the visceral epithelium constructed of podocytes (see text figures 25a and b).

The surrounding tissue of the kidney cortex comprises three kinds of vessels or tubules. Of these the proximal convoluted tubules (PCT) are the most prominent. They have relatively thick walls, the inner surfaces of which are covered with microvilli (Plate 24). At this magnification the microvilli are not resolved, and the only textural component of this surface that can be seen appears as small black dots, looking like pores, in the wall. Actually these are holes or breaks in the otherwise close packing of microvilli. The distal tubles (DT) are much less numerous and have thinner walls. Their luminal surfaces appear much smoother. The blood vascular elements between the tubules of the nephrons are so small as to be almost indiscernible at this magnification. They are mostly the size and nature of capillaries (Cp) with very thin walls.

A break in the tissue, produced by the cutting, is shown at the asterisk (*).

From the rat
Magnification × 500

References

BERGER, S. J., and SACKTOR, B. Isolation and biochemical characterization of brush borders from rabbit kidney. *J. Cell Biol., 47*:637 (1970).

BOURDEAU, J. E., CARONE, F. A., and GANOTE, C. E. Serum albumin uptake in isolated perfused renal tubules. *J. Cell Biol., 54*:382 (1972).

BULGER, R. E. The shape of rat kidney tubular cells. *Amer. J. Anat., 116*:237 (1965).

DIAMOND, J. M. Transport of salt and water in rabbit and guinea pig gall bladder. *J. Gen. Physiol., 48*:1 (1964).

——, and TORMEY, J. McD. Role of long extracellular channels in fluid transport across epithelia. *Nature, 210*: 817 (1966).

GIEBISCH, G., BOULPAEP, E. L., and WHITTEMBURY, G. Electrolyte transport in kidney tubule cells. *Phil. Trans. Roy. Soc. London, Ser. B., 262*:175 (1971).

GRAHAM, R. C., JR., and KARNOVSKY, M. J. The early stages of absorption of injected horseradish peroxidase in the proximal tubules of mouse kidney: ultrastructural cytochemistry by a new technique. *J. Histochem. Cytochem. 14*:291 (1966).

HEIDRICH, H.-G., KINNE, R., KINNE-SAFRAN, E., and HANNIG, K. The polarity of the proximal tubule cell in rat kidney. *J. Cell Biol., 54*:232 (1972).

LATTA, H., MAUNSBACH, A. B., and OSVALDO, L. The fine structure of renal tubules in cortex and medulla. *In* Ultrastructure of the Kidney. A. J. Dalton and F. Haguenau, editors. New York, Academic Press (1967) p. 1.

MAUNSBACH, A. B. Functions of lysosomes in kidney cells. *In* Lysosomes in Biology and Medicine. Volume I. J. T. Dingle and H. B. Fell, editors. Amsterdam, North Holland Publishing Co. (1969) p. 115.

MILLER, F. Hemoglobin absorption by the cells of the proximal convoluted tubule in mouse kidney. *J. Biophysic. and Biochem. Cytol., 8*:689 (1960).

PEASE, D. C. Electron microscopy of the tubular cells of the kidney cortex. *Anat. Rec., 121*:721 (1955).

RHODIN, J. Anatomy of kidney tubules. *Int. Rev. Cytol., 7*:485 (1958).

SOLOMON, A. K. Pumps in the living cell. *Sci. Amer., 207*:100 (August, 1962).

STRAUS, W. Cytochemical observations on the relationship between lysosomes and phagosomes in kidney and liver by combined staining for acid phosphatase and intravenously injected horse-radish peroxidase. *J. Cell. Biol., 20*:497 (1964).

TRUMP, B. F., and BULGER, R. E. Morphology of the kidney. *In* Structural Basis of Renal Disease. E. L. Becker, editor. New York, Hoeber Medical Division, Harper and Row (1968) p. 1.

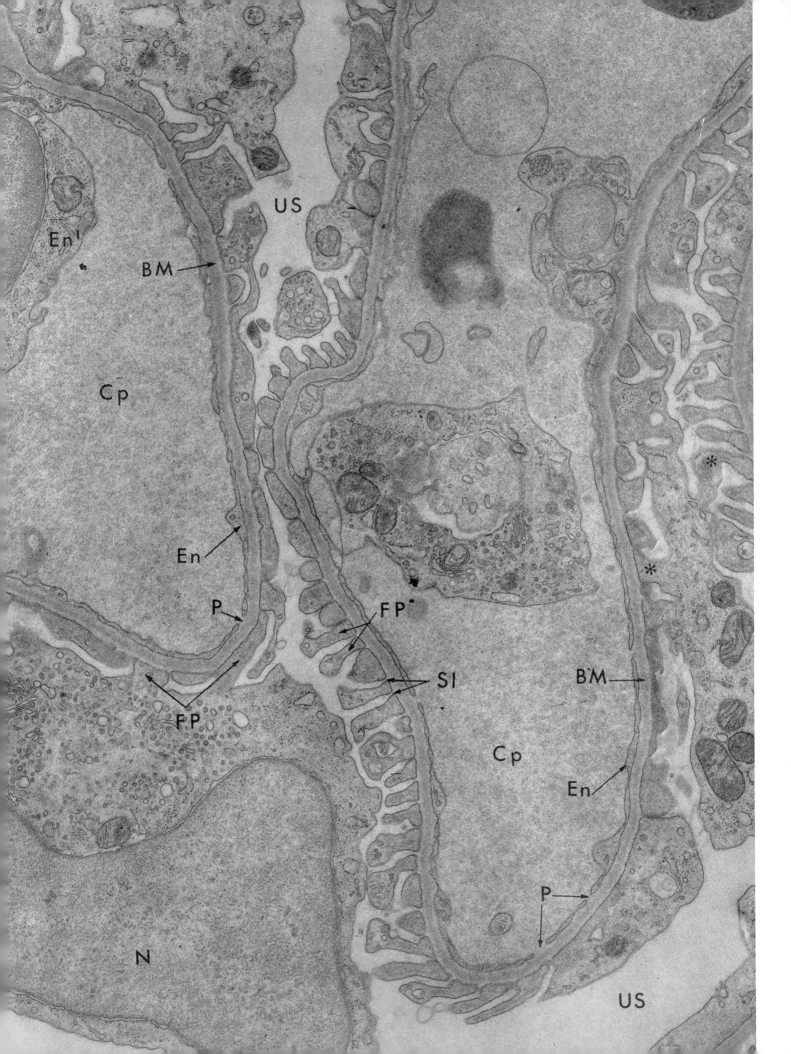

PLATE 25

The Renal Corpuscle

In the corpuscle of the kidney, urine is formed initially as an ultrafiltrate of blood: molecules with a molecular weight above approximately 45,000 are retained in the circulatory system, while smaller ones filter into the urinary space or lumen of the nephron (see Plate 24). In this electron micrograph the morphological nature of the renal filter can be examined.

The capillary vessels that make up the glomerulus are lined by extremely thin endothelial cells (En). Only in the region of its nucleus (as at En′) does the cell project into the lumen of the vessel (Cp). For the most part, the capillary endothelial cells in this micrograph are depicted in the profile of their smallest dimension. Thus can be visualized one of the most salient features of these cells: the pores or fenestrae (P), which place the blood plasma in direct contact with the underlying basement membrane (BM). In the corpuscle the latter is relatively thick, representing as it does the fusion of two such membranes, that of the glomerular endothelium and that of the capsular epithelium. Recently, partly on the basis of EM studies, another cell type, the "deep" cell, has been identified. It is usually interposed between the endothelial cells and the basement membrane.

The epithelium, which lines the urinary space of Bowman's capsule, has also been shown to possess some unusual features. It is composed of flat cells, which send out small cytoplasmic processes or "feet" that rest on the basement membrane. These processes interdigitate so extensively with those of adjacent cells that in almost any section cut normal to the basement membrane, the epithelium appears as a series of footlike profiles. As a result of this morphology these cells have been named podocytes. In this micrograph the nucleus (N) of a podocyte is seen, and the cytoplasmic extensions (FP) of this cell as well as those of other podocytes can be observed resting against the basement membrane (BM). It is known that individual podocytes may have "feet" implanted upon basement membranes of several adjacent capillaries, and indeed in this picture the "feet" of one cell extend to two different capillaries (*). In this intricate arrangement of the epithelial cells of Bowman's capsule, spaces or slits (Sl) are left between the cytoplasmic "feet." However, in preparations of exceptional quality these slits appear to be traversed by extremely thin diaphragms. These, together with the basement membrane, would therefore be regarded as the renal filter that determines which molecular species can pass into the nephron. Studies of nephrotic kidneys indicate that the properties of this filter are monitored by the adjacent cell layers. In this activity the endothelial cells, especially the "deep" cells, show increased phagocytosis of injected tracer molecules such as ferritin. An even more marked response to such challenges has been observed in the podocytes. Fewer slits between the "feet" are found, and those that remain often seem to be closed by junctions that prevent passage of material. Rather than leaking through the slits, the marker is incorporated into various vacuoles, granules, and bodies within the epithelial cells and must cross this cellular barrier before reaching the urinary space (US). The cellular responses in nephrosis are believed to be part of a mechanism that can compensate, though insufficiently, for increased and abnormal permeability of the filter.

From the kidney of the mouse
Magnification × 29,500

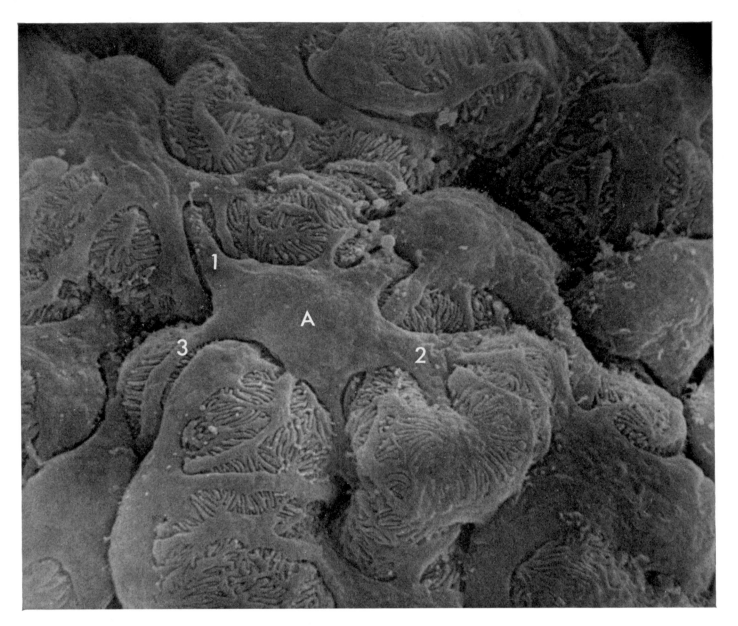

Text Figure 25a

The podocyte is one of the most remarkable cells in the body. In isolation it would appear as a cell with several arms, each with many branches and sub-branches. The least of these exist as foot processes (pedicles) that *in situ* interdigitate with identical processes of other podocytes to form a continuous epithelium over the visceral surface of the glomerular capillaries. These features of the podocyte are depicted in this scanning electron micrograph. The cell at A extends processes to at least three (1,2,3) capillaries and possibly more. At their extremities the major arms of the podocytes break up into smaller branches that in turn subdivide into foot processes. Where the latter interdigitate with others of adjacent podocytes, an intercellular contact of considerable length is generated. This length can be quite simply varied (shortened), as it is in some forms of nephritis, by a shortening of the foot processes and an expansion of the smaller branches and arms. Thus the effective filter (the slits between the pedicles) is decreased.

This micrograph illustrates the extraordinary opportunities offered by scanning microscopy for studying the pathologic states of this physiologically very important epithelium.

From the rat
Magnification × 4,600

110

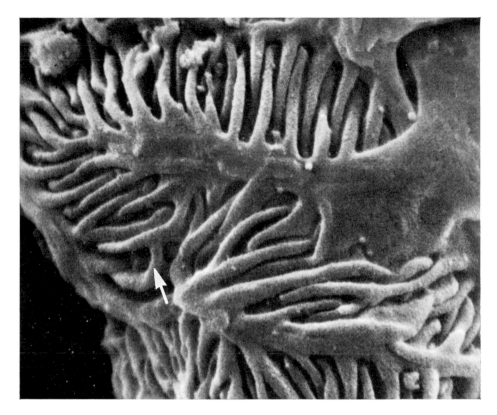

Text Figure 25b

In this scanning electron micrograph the details of the association among podocytes, already described in the legends of Plate 25 and text figure 25a, become strikingly evident. In its simplest form this mingling of pedicles is shown at the bottom of the micrograph. The outlines of individual pedicles are clear. The dark zone of separation between them is deceptive in that the actual separation at the level of the basal lamina is too fine for scanning resolution, even if it were not in shadow. The individual foot processes flatten out laterally as they contact the underlying surface of the lamina (see Plate 25).

In other parts of the micrograph the processes of one podocyte cross over those of another (arrow) and in some instances seem to rest on the surface of another. Although not evident here, bridges between pedicles have been noted in some instances, and these may represent the regions of gap junctions to be expected in any epithelium. They have been noted especially where the epithelium is altered, as in nephritis.

From the rat
Magnification × 14,000

References

ARAKAWA, M., and TOKUNAGU, J. A SEM study of the glomerulus: further consideration of the mechanism of the fusion of podocyte terminal processes in nephrotic rats. *Lab. Invest.*, 27:366 (1972).

BUSS, H. Development of podocytes in the rat renal glomerulus. A SEM study. *In* Microscopie Électronique 1970 Volume III. P. Favard, editor. Paris, Société Française de Microscopie Électronique (1970) p. 605.

———, LAMBERTS, B., and BRASS, H. Orthology and pathology of the renal podocytes. *In* Scanning Electron Microscopy/1973. O. Johari and I. Corvin, editors. Chicago, IIT Research Institute (1973) p. 573.

FARQUHAR, M. G., and PALADE, G. E. Glomerular permeability. II. Ferritin transfer across the glomerular capillary wall in nephrotic rats. *J. Exp. Med.*, 114:699 (1961).

———, WISSIG, S. L., and PALADE, G. E. Glomerular permeability. I. Ferritin transfer across the normal glomerular capillary wall. *J. Exp. Med.*, 113:47 (1961).

KURTZ, S. M., and FELDMAN, J. D. Experimental studies on the formation of the glomerular basement membrane. *J. Ultrastruct. Res.*, 6:19 (1962).

LATTA, H., MAUNSBACH, A. B., and MADDEN, S. C. The centrolobular region of the renal glomerulus studied by electron microscopy. *J. Ultrastruct. Res.*, 4:455 (1960).

MENEFEE, M. G., and MUELLER, C. B. Some morphological considerations of transport in the glomerulus. *In* Ultrastructure of the Kidney. A. J. Dalton and F. Haguenau, editors. New York, Academic Press (1967) p. 73.

RHODIN, J. A. G. The diaphragm of capillary endothelial fenestrations. *J. Ultrastruct. Res.*, 6:171 (1962).

TYLER, W. S., DUNGWORTH, D. L., and NOWELL, J. A. The potential of SEM in studies of experimental and spontaneous diseases. *In* Scanning Electron Microscopy/1973. O. Johari and I. Corvin, editors. Chicago, IIT Research Institute (1973) p. 404.

YAMADA, E. The fine structure of the renal glomerulus of the mouse. *J. Biophysic. and Biochem. Cytol.*, 1:551 (1955).

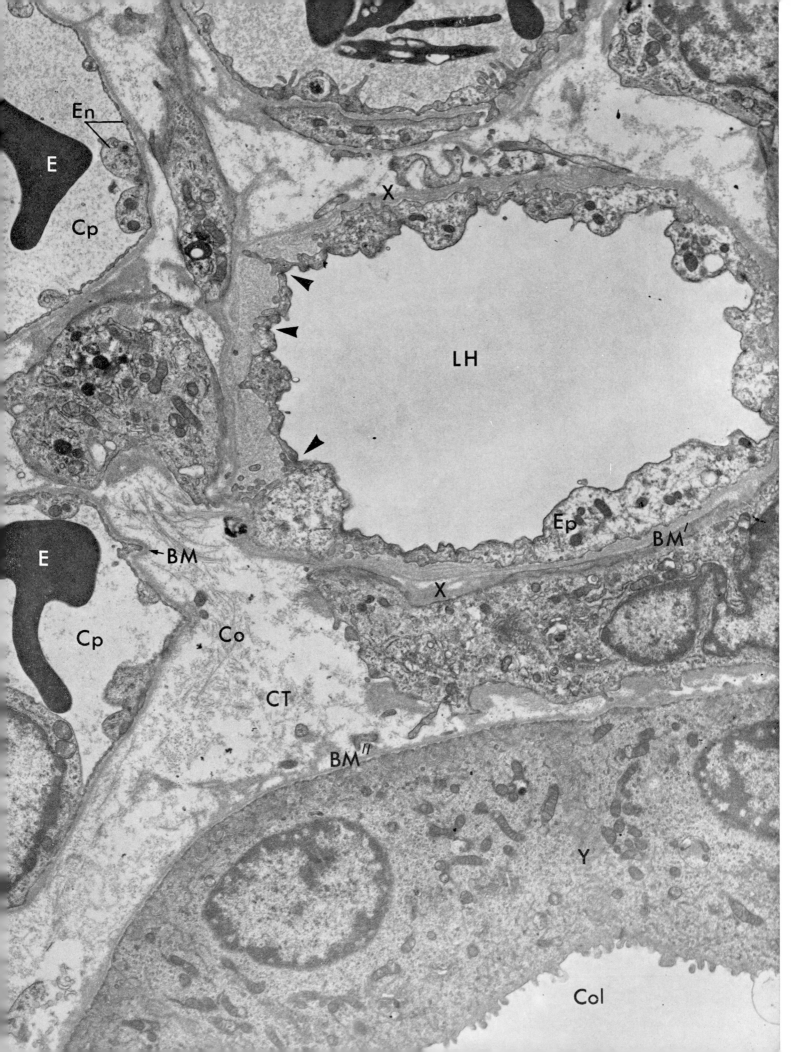

PLATE 26

The Renal Medulla

IN birds and mammals the nephrons of the kidney each include a segment that bends back on itself and is called the loop of Henle (LH). The filtrate formed in the glomerulus (Plate 25) is reduced in volume by reabsorption in the proximal tubule (Plate 24) and then enters the descending limb of this segment. The adjective *descending* is used to indicate that the tubule extends from the cortical region of the kidney into the deeper inner region called the medulla. The other arm of the loop, the ascending limb, extends toward the cortex and leads into the thicker walled distal tubule. The two arms of the loop lie near each other, separated by connective tissue, the interstitium (CT) of the renal medulla. In addition, the medulla contains numerous collecting tubules (Col) that open into the renal pelvis. As their name suggests, each of the latter tubules drains urine from a number of nephrons.

The length of the loop of Henle varies from species to species, and it was noted over fifty years ago that its length shows a positive correlation with the ability of the kidney to produce a concentrated urine. Thus it was realized that the loops must contribute to the vital functioning of the kidney, effecting elimination of waste metabolites while conserving water needed especially by land-dwelling vertebrates.

It is only relatively recently, after painstaking study, that the way in which the loops function to accomplish this task has begun to be understood. First it was important to learn by freezing-point determinations, both of the filtrate within the nephrons and of the extracellular fluid, that as samples are taken at levels closer to the tips of the loops and therefore nearer to the tip of the renal papilla, the concentration of salt increases both in the loops and in the extracellular fluids around them. Thus there exists in the medulla a steep gradient of salt concentration, which is thought to be maintained by a salt-pumping segment of the loop. As pointed out above, the collecting tubules also traverse this region and extend toward the renal pelvis. They therefore pass through a zone in which the surrounding medium is increasingly hypertonic. As a result, water moves out of the tubules into the interstitium, and the urine becomes more concentrated. This interesting device for conserving water and salts for the animal's repeated use while simultaneously eliminating metabolic wastes is considered further below.

Before describing loop function in more detail, however, we should discuss the structure and arrangement of the loop and other medullary tissues. Cross sections of the thin segments of the loops of Henle (LH) differ significantly from adjacent capillaries (Cp) lying near them. The simple squamous epithelium (Ep) lining the loops is slightly thicker than the fenestrated capillary endothelium (En) (which in places is extremely attenuated), and its surface may be somewhat irregular, with small projections extending into the lumens. Its basal surface is also fairly uneven. Since adjacent cells in this epithelium interdigitate in a complex manner, the micrograph shows numerous junctions (arrows) between them. No evidence of pinocytosis has been observed in the tubular epithelium, although machinery for such activity is frequently present in endothelia. The presence in the capillary lumens (Cp) of erythrocytes (E) and coagulated plasma proteins contrasts with the apparently empty thin segments. A basement membrane underlies both the capillary endothelium (BM) and the tubular epithelium (BM'). In the latter, however, this layer is unusually thick and often consists of several lamellae (X). It may be noted in passing that the interstitial spaces contain remarkably little collagen (Co) but an abundance of basement membrane material and ground substance.

As it reaches the outer zone of the renal medulla (nearer the cortex) the ascending limb of the loop of Henle becomes thicker. This thicker portion joins the distal convoluted tubule, which lies within the kidney cortex. Although the thick region of the loop resembles the distal tubule in structure, its special functions as part of the loop have not yet been distinguished physiologically. Therefore it is treated with the loop as a whole in the discussion of loop function given below.

The collecting tubules (Col) are lined by cuboidal epithelial cells. Their basal surfaces, resting on a thin basement membrane (BM"), are highly folded, although the folds are much less deep than those of the cells in the proximal tubule (see Plate 24). Their lateral surfaces are also highly folded and interdigitate with folds of adjacent cells (Y). Irregular, short, villuslike projections are found on their free surfaces. The transport of water through the walls of the collecting tubules is under control of the antidiuretic hormone (ADH), a polypeptide secreted via the neuro-hypophysis. Within certain limits more water may be conserved and the urine made more concentrated as the hormone acts to make the epithelium increasingly permeable to water.

We can now return to a consideration of how the loops of Henle bring about a hypertonic condition in the interstitium. To account for this function it is assumed that the ascending limbs actively pump out sodium into the extracellular spaces and that they themselves are impermeable to water. Thus filtrate flowing in the descending limbs progressively loses water to the hypertonic interstitium and becomes more concentrated. On entering the ascending limb, sodium is actively removed, and, although the amount is relatively small at any one point along the limb (because the difference in solute concentration between the inside and the outside of the ascending limb is never greater than ~ 200 mOsm/L), it is enough to maintain the gradient. Water, however, is unable to penetrate this side of the loop. The filtrate in losing salt but not water therefore decreases in tonicity and finally becomes hypotonic before reaching the distal tubule. Thus the hairpin-like tubules constitute a kind of countercurrent multiplier system. It is the nature of such a system that the longer the loops the more concentrated will be the tubular filtrate at the point where the tubule turns back on itself. In the kidney, it would follow that the longer the loops, the more hypertonic the interstitium at the tip of the renal papilla and thus the more concentrated will the urine become because of water removal as it flows through the collecting tubules.

The capillaries of the medulla are also arranged in loops and constitute a second countercurrent multiplier system. They are intimately associated with the loops of Henle, as shown in this image. It is a result of this arrangement that the hyper-tonic gradient of the medulla is maintained while water and salt are being removed from the kidney. The blood, flowing slowly into descending capillary loops, loses water and gains solutes, chiefly sodium, as the gradient of the interstitium increases around it. At the apex of the loop, deep in the medulla, the tonicity of the blood equilibrates with the surrounding medium, so that the blood as well as the urine becomes concentrated. The crenated appearance of the erythrocytes in this micrograph attests to the hypertonicity of the plasma. However, instead of leaving the kidney at the papilla, the capillary bends back upon itself. Thus on its course toward the renal cortex, the blood loses solutes and regains water without markedly disturbing the gradient. Any salt and water carried away in excess of that brought to the medulla by the capillaries are compensated for by the pumping action of the ascending limbs mentioned above.

Satisfactory preparation of the renal medulla for electron microscopy has proved to be difficult. This is doubtless due to the impossibility of finding a single fixative suitable to the varied tonicities in the tissue. Those who have studied its fine structure, however, have been somewhat dismayed to discover the "simple" structure, especially of the thin segment of the ascending loops, which are thought to be highly active physiologically. Indeed, when the two thin limbs have been identified with certainty in thin sections, even the epithelium of the descending limb appears more complex in structure. The correlation between structure and function in the case of the loop of Henle is therefore less than satisfactory. To some investigators this finding has suggested that the countercurrent theory of loop function may require modification. To others, it seems likely that the morphologists' concept of the fine structural features necessarily associated with actively transporting epithelia should be modified.

From the kidney of the rat
Magnification $\times$ 11,000

References

BERLINER, R. W. Some aspects of the function of the renal medulla. *In* Progress in Pyelonephritis. E. H. Kass, editor. Philadelphia, F. A. Davis Company (1965) p. 417.

BULGER, R. E., TISHER, C. C., MYERS, C. H., and TRUMP, B. F. Human renal ultrastructure. II. The thin limb of Henle's loop and the interstitium in healthy individuals. *Lab. Invest., 16:*124 (1967).

GANOTE, C. E., GRANTHAM, J. J., MOSES, H. L., BURG, M. B., and ORLOFF, J. Ultrastructural studies of vasopressin effect on isolated perfused renal collecting tubules of the rabbit. *J. Cell Biol., 36:*355 (1968).

GOTTSCHALK, C. W. Osmotic concentration and dilution of the urine. *Amer. J. Med., 36:*670 (1964).

GRANTHAM, J. J., GANOTE, C. E., BURG, M. B., and ORLOFF, J. Paths of transtubular water flow in isolated renal collecting tubules. *J. Cell Biol., 41:*562 (1969).

MARSH, D. J., and SOLOMON, S. Analysis of electrolyte movement in thin Henle's loops of hamster papilla. *Amer. J. Physiol., 208:*1119 (1965).

OSVALDO, L., and LATTA, H. The thin limbs of the loop of Henle. *J. Ultrastruct. Res., 15:*144 (1966).

PITTS, R. F. Physiology of the Kidney and Body Fluids. Chicago, Year Book Medical Publishers Incorporated (1963).

SCHMIDT-NIELSEN, B., and O'DELL, R. Structure and concentrating mechanism in the mammalian kidney. *Amer. J. Physiol., 200:*1119 (1961).

ULLRICH, K. J., and MARSH, D. J. Kidney, water, and electrolyte metabolism. *Ann. Rev. Physiol., 25:*91 (1963).

YOUNG, D., and WISSIG, S. L. A histologic description of certain epithelial and vascular structures in the kidney of the normal rat. *Amer. J. Anat., 115:*43 (1964).

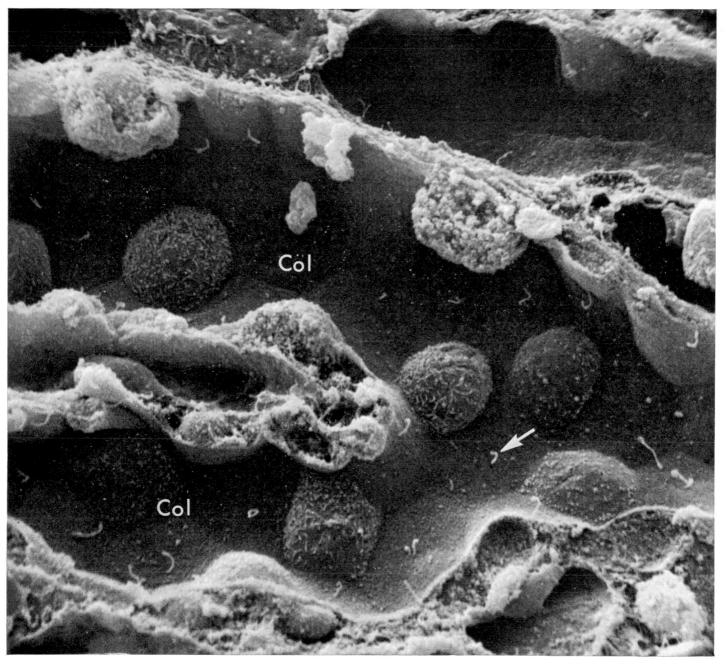

Text Figure 26a

After fixation by vascular perfusion, tissue from the renal medulla was torn open to reveal the surface of the epithelium lining the tubules and vesicles of this area. In this scanning electron micrograph, two collecting tubules (Col) join together. Two types of configurations are apparent. In one type the polygonal surface faces of the cells have a generally smooth appearance with a few very short projections; a cilium (arrow), which has been identified as such in transmission micrographs, projects from the center of each cell. In the other type the cells are seen as round, shallow domes, scattered individually among the flatter cells and rising higher from the surface.

In this type there is an array of rather short microvilli. There is no hypothesis to explain the significance of the two types. The damaged remnants of epithelial cells are sometimes seen at the torn edges of the tubule.

At the top of the micrograph is another type of tubule, perhaps part of the loop of Henle. The intimacy of adjacent medullary tubules is apparent and emphasizes the sparse connective tissue of the organ, as can be seen also in transmission electron micrographs (Plate 26).

From the rat
Magnification × 2,600

115

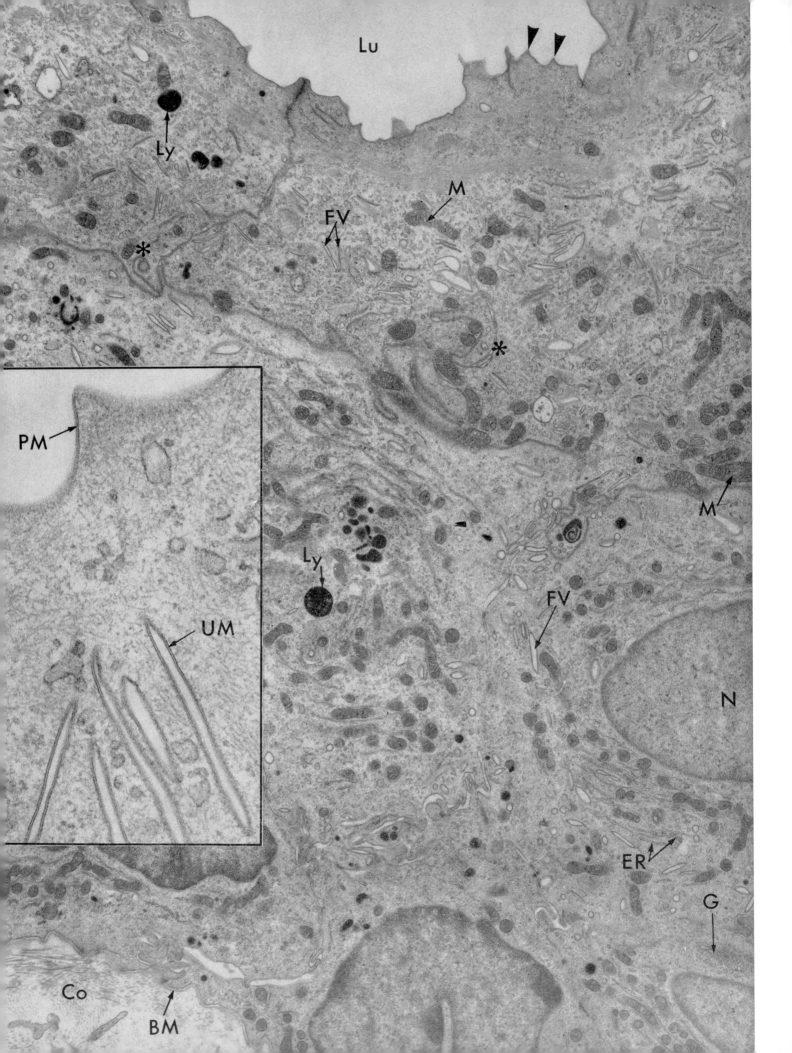

Lu

Ly

*

FV

M

*

M

PM

Ly

FV

UM

N

ER

Co

G

BM

PLATE 27

The Transitional Epithelium

THE urinary tract is lined by a transitional epithelium, which is remarkable for the rapidity with which it can accommodate to changes in the area of the surface it must cover. Examination of the fine structure of this tissue affords some insight into its mode of function.

The epithelium depicted in this micrograph consists of two to three layers of cells resting on a thin basement membrane (BM) and an underlying layer of collagen fibrils (Co). As in most cell types, nuclei (N), mitochondria (M), lysosomes (Ly), and extensive Golgi regions (G), as well as rough-surfaced cisternae of the endoplasmic reticulum (ER), can readily be identified. In addition, certain structural features distinguish the cells of this tissue. For example, the surface bordering the lumen (Lu) of the bladder is marked by crests and hollows, giving it a scalloped appearance (arrows, upper right). The cell cortex just beneath this surface is apparently reinforced by a thick feltwork of finely fibrous material. Within this and also in the deeper cytoplasm of the superficial cells there are many vesicles rendered peculiar by their fusiform profiles (FV). Even at low magnifications the membrane limiting these vesicles seems thick, but when examined more closely at higher magnifications (inset), it is found to have the dimensions (120 A) and the trilaminar unit membrane structure (UM) identical to those of the plasma membrane (PM) that covers the cell's free surface. Furthermore, the structure and dimensions of clefts in the free surface suggest that the compressed vesicles form when opposing wave crests fuse and the troughs between them are pinched off into the cytoplasm.

That this in fact happens has been demonstrated by injecting a marker such as ferritin into the lumen of the bladder and following its uptake into vesicles such as those shown in this micrograph (FV and inset). This activity is clearly evident when the bladder contracts and forms deep clefts in its otherwise smooth surface. The surface membrane thus incorporated into the cell seems to find its way eventually into lysosomes, where it is presumably destroyed.

The purpose for this seemingly extravagant use of this surface membrane is not immediately evident. However, it has been pointed out that this epithelium requires a special surface membrane to function as a barrier to the free diffusion of water, for otherwise the hypertonic urine would draw water from the tissue. When, in fact, the membrane is damaged experimentally by proteolytic agents, lipid solvents, or sodium thioglycolate and its continuity thereby disrupted, the permeability of the epithelium to water is found to increase approximately tenfold. It follows as reasonable, therefore, that if the surface membrane wears out or fails to perform its function adequately, it should be removed from the surface and broken down by lysosomes (Ly). Constant replacement of surface membrane would consequently be required, and some evidence is available that the Golgi complex here is the site of synthesis of new cell surface.

A further complexity of structure is illustrated by the numerous interfoldings (*) and interdigitations that mark the contacts between the deeper cells of this epithelium. It seems that these lateral folds represent a kind of membrane "storage" and that when the bladder is distended the folds disappear to give relatively smooth cell surfaces. It is noteworthy too that only a few small desmosomes connect these epithelial cells, so that they can slide more freely over one another as the bladder changes its volume.

From the urinary bladder of the mouse
Magnification × 8,500
Inset × 77,000

References

CHLAPOWSKI, F. J., BONNEVILLE, M. A., and STAEHELIN, L. A. Lumenal plasma membrane of the urinary bladder. II. Isolation and structure of membrane components. *J. Cell Biol.*, 53:92 (1972).

HICKS, R. M. The fine structure of the transitional epithelium of rat ureter. *J. Cell Biol.*, 26:25 (1965).

———. The permeability of rat transitional epithelium. Keratinization and the barrier to water. *J. Cell Biol.*, 28:21 (1966).

———. The function of the Golgi complex in transitional epithelium. Synthesis of the thick cell membrane. *J. Cell Biol.*, 30:623 (1966).

———, and KETTERER, B. Isolation of the plasma membrane of the lumenal surface of rat bladder epithelium, and the occurrence of a hexagonal lattice of subunits both in negatively stained whole mounts and in sectioned membranes. *J. Cell Biol.*, 45:542 (1970).

STAEHELIN, L. A., CHLAPOWSKI, F. J., and BONNEVILLE, M. A. Lumenal plasma membrane of the urinary bladder. I. Three-dimensional reconstruction from freeze-etch images. *J. Cell Biol.*, 53:73 (1972).

VERGARA, J., LONGLEY, W., and ROBERTSON, J. D. A hexagonal arrangement of subunits in membrane of mouse urinary bladder. *J. Mol. Biol.*, 46:593 (1969).

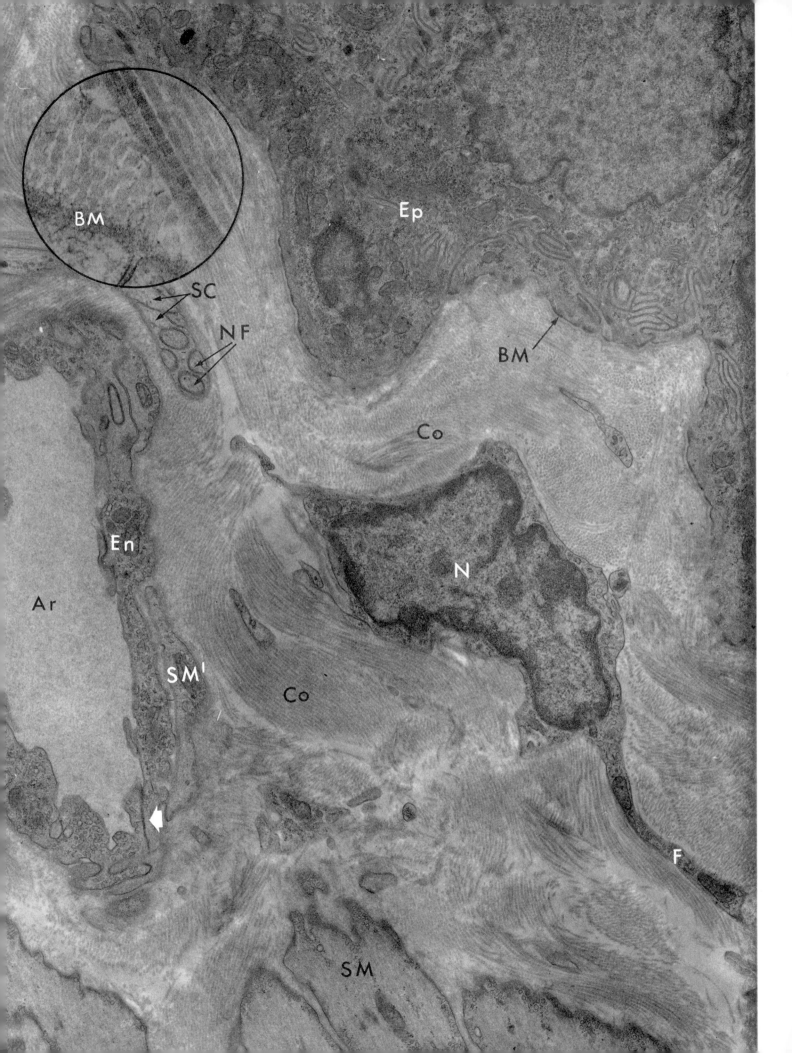

PLATE 28

Connective Tissue of the Lamina Propria

CONNECTIVE tissue is defined most simply as that tissue in which there is a substantial amount of intercellular matrix relative to the number of cells present. It is the matrix that is of incomparable importance in the structuring of the vertebrate body, for it binds together, supports, and protects other types of tissue. This vast intercellular material is made up of protein fibers embedded in a continuous ground substance consisting mainly of mucopolysaccharides. Variations in the proportions of fibers to ground substance, as well as chemical differences among the fiber types and among the kinds of mucopolysaccharides present, account for some of the differences observed among the several types of connective tissue. These vary from the lubricating synovial fluid of the joints to the pliant layers supporting epithelia, to the rigid rods and plates of cartilage and bone, and to the ropelike tendons. In each type of connective tissue, both the fibers and ground substance seem to be formed by a single cell type, which therefore determines many of the properties of the tissue.

The lamina propria, a delicate layer of subepithelial connective tissue, will serve to illustrate a number of properties shared generally by connective tissues. A feltwork of collagen fibril bundles (Co) is the chief supporting element in this example. These long, flexible, but nonelastic fibers are embedded, as mentioned above, in an amorphous ground substance, which appears in the micrograph as a material of low density between the fibrils. When the matrix is forming rapidly, the fibroblasts that produce it have abundant cytoplasm, rich in rough-surfaced ER and possessing a large Golgi region. However, after they have laid down this matrix, the fibroblasts, or fibrocytes as they should then be called, become quiescent, and evidence of synthesis and secretion is no longer detected. The fibrocyte nucleus (N) is then surrounded by only a shallow layer of cytoplasm, which extends out into thin, flattened processes (F).

The lamina propria binds together the esophageal epithelium (Ep) and the smooth muscle (SM) of the muscularis mucosae. In addition it supports blood vessels and nerve fibers. A small precapillary arteriole (Ar) is included at the left. It has a lumen enclosed by endothelial cells (En)

that have no pores or fenestrae and are closely joined at their margins (arrow; see also Plate 10). As in the case of other endothelial cells (Plates 10 and 39), small pinocytotic vesicles are frequently associated with their plasma membranes, and many small vesicles can be seen throughout the endothelial cytoplasm. Unlike capillaries proper, this precapillary vessel has a few isolated smooth muscle elements associated with it (SM'). The wavy outline of the endothelium, typical of arterial vessels in histological preparations, is a result of the contraction of this muscle layer (see also Plates 40 and 41). Cross sections of a few nerve fibers (NF) enclosed in Schwann cell cytoplasm (SC) can be identified near the blood vessel, but further discussion of these is deferred to the legend of Plate 44.

Usually, the connective tissue harbors a number of other cell types. Most of them—for example, macrophages, plasma cells, wandering leukocytes, and mast cells—aid in resisting invasion by foreign organisms or harmful chemicals.

Another extracellular structure deserves discussion at this point. The connective tissue matrix is separated from other tissues by a thin sheet of amorphous material, the basement membrane or, as it is now often called, the basal lamina. It appears here as a thin line (BM) underlying the epithelium and may also be found enclosing the blood vessels, the Schwann cells that surround nerve fibers, and the muscle cells. The membrane, which on closer examination appears to be made up of fine filaments randomly oriented (see inset, BM), is argyrophilic and is rich in mucopolysaccharide. Only matrix-producing cells, such as fibroblasts, and mobile cells that move through the connective tissue lack this continuous membranous covering.

The basal lamina is always intimately associated with collagen fibrils. On its side facing the connective tissue, for example, the lamina material seems to fray out and intermingle with collagen fibrils so that some of the fibrils are anchored in the basement membrane. This relationship undoubtedly serves to bind connective tissues and epithelia together. The basal lamina may also act as a kind of molecular sieve influencing exchange of metabolites between the matrix and the cells it encloses (cf. Plate 25) and restraining the flow of larger molecular species, e.g., 50,000 mol. wt. Some impression of the substructure of the basal lamina may be seen in lower half of the inset (BM′) where the layer, as it occurs under capillary endothelial cells, blends with collagen fibrils.

The basal lamina is now generally believed to originate from the epithelial or other cell it underlies and not from the connective tissue. Identification within epithelial cells of presumptive precursors having the same antigenic and staining properties as the basal lamina has led to this conclusion.

Most collagen fibers, as opposed to the basal lamina, clearly owe their existence to fibroblasts. When these cells are isolated in tissue culture, fibrils, singly or in bundles, appear first in close association with the cell surface. Thin extracellular protofibrils subsequently increase in size by accretion of units of collagen, the tropocollagen molecules, from the tissue ground substance. The banded pattern of collagen fibrils (see inset) reflects the orderly assemblage of the highly asymmetric macromolecular units into fibrils of indefinite length. From pieces of tendon, which are largely collagen, it is possible to isolate the fibrous protein in pure form, and most of what is known about its architecture and properties has been gleaned from this material.

The ground substance, too, has been investigated with great care. Once its existence was demonstrated, its origin from fibroblasts in tissue and organ cultures was detected. In regard to its composition, the work of Karl Meyer and his colleagues has contributed most to the characterization of the polysaccharides found in a variety of connective tissues. These molecules form a viscous molecular mesh, a medium through which cells and metabolites must travel. The properties, as well as the origins of such a ground substance, will be considered in further detail in relation to Plate 29.

Because both fibers and ground substance are produced by the same cell, the cytological events associated with collagen synthesis are difficult to separate from those concerned with polysaccharide formation. Electron microscope studies of growing connective tissue supplied with radioactive proline (an amino acid abundant in collagen) indicate that the collagen precursors may first be found in the cisternae of the rough-surfaced endoplasmic reticulum. The manner in which these molecules leave the cell and the involvement of the cell in bringing about their final aggregation to form collagen fibrils are currently debated by various workers in the field. Some, on the one hand, think that the Golgi region has a role in the elaboration of tropocollagen that is possibly analogous to its role in the formation of zymogen granules by the exocrine pancreas (see Plate 11). Others claim, however, that collagen fibril precursors may bypass the Golgi complex and that the ER cisternae may open directly into the extracellular space to release tropocollagen. In this latter interpretation the activity of the Golgi complex would be correlated largely with the synthesis of ground substance (see Plate 29).

From the esophagus of the bat
Magnification × 15,000
Inset × 77,000

References

CONRAD, G. W. Collagen and mucopolysaccharide biosynthesis in the developing chick cornea. *Dev. Biol.,* *21:*292 (1970).
———. Collagen and mucopolysaccharide biosynthesis in mass cultures and clones of chick corneal fibroblasts *in vitro. Dev. Biol., 21:*611 (1970).
FILTON JACKSON, S. The morphogenesis of collagen. *In* Treatise in Collagen 2 (pt. B). B. S. Gould, editor. New York, Academic Press (1968) p. 1.
GROSS, J. Collagen. *Sci. Amer., 204:*120 (May, 1961).
HAY, E. D. Organization and fine structure of epithelium and mesenchyme in the developing chick embryo. *In* Epithelial-Mesenchymal Interactions. Baltimore, Williams and Wilkins (1968) p. 31.
———, and DODSON, J. W. Secretion of collagen by corneal epithelium. I. Morphology of the collagenous products produced by isolated epithelia grown on frozen-killed lens. *J. Cell Biol., 57:*190 (1973).
HODGE, A. J., and SCHMITT, F. O. The tropocollagen macromolecule and its properties of ordered interaction. *In* Macromolecular Complexes. M. V. Edds, Jr., editor. New York, The Ronald Press Co. (1961) p. 19.
KARRER, H. E. Electron microscope study of developing chick embryo aorta. *J. Ultrastruct. Res., 4:*420 (1960).
MANASEK, F. J. Sulfated extracellular matrix production in the embryonic heart and adjacent tissues. *J. Exp. Zool., 174:*415 (1970).
NADOL, J. B., JR., and GIBBINS, J. R. Autoradiographic evidence for epithelial origin of glucose-rich components of the basement membrane (basal lamina) and basement lamella in the skin of *Fundulus heteroclitus. Z. Zellforsch. Mikrosk. Anat., 106:*398 (1970).
———, GIBBINS, J. R., and PORTER, K. R. A reinterpretation of the structure and development of the basement lamella: an ordered array of collagen in fish skin. *Dev. Biol., 20:*304 (1969).
PIERCE, G. B., BEALS, T. F., RAM, J. S., and MIDGLEY, A. R. Basement membranes. IV. Epithelial origin and

immunologic cross reactions. *Amer. J. Path., 45:*929 (1964).

PORTER, K. R. Cell fine structure and biosynthesis of intercellular macromolecules. *Biophys. J., 4* (Suppl.): 167 (1964).

————, and PAPPAS, G. D. Collagen formation by fibroblasts of the chick embryo dermis. *J. Biophysic. and Biochem. Cytol., 5:*153 (1959).

ROSS, R., and BENDITT, E. P. Wound healing and collagen formation. III. A quantitative radiographic study of the utilization of proline-H³ in wounds from normal and scorbutic guinea pigs. *J. Cell Biol., 15:*99 (1962).

————, and BENDITT, E. P. Wound healing and collagen formation. V. Quantitative electron microscope radio-autographic observations of proline-H³ utilization by fibroblasts. *J. Cell Biol., 27:*83 (1965).

STARK, M., MILLER, E. J., and KÜHN, K. Comparative electron microscope studies on the collagens extracted from cartilage, bone, and skin. *Eur. J. Biochem., 27:* 192 (1972).

TRELSTAD, R. L., and COULOMBRE, A. J. Morphogenesis of the collagenous stroma in the chick cornea. *J. Cell Biol., 50:*840 (1971).

YARDLEY, J. H., HEATON, M. W., GAINES, L. M., JR., and SHULMAN, L. E. Collagen formation by fibroblasts. *Bull. Hopkins Hosp., 106:*381 (1960).

121

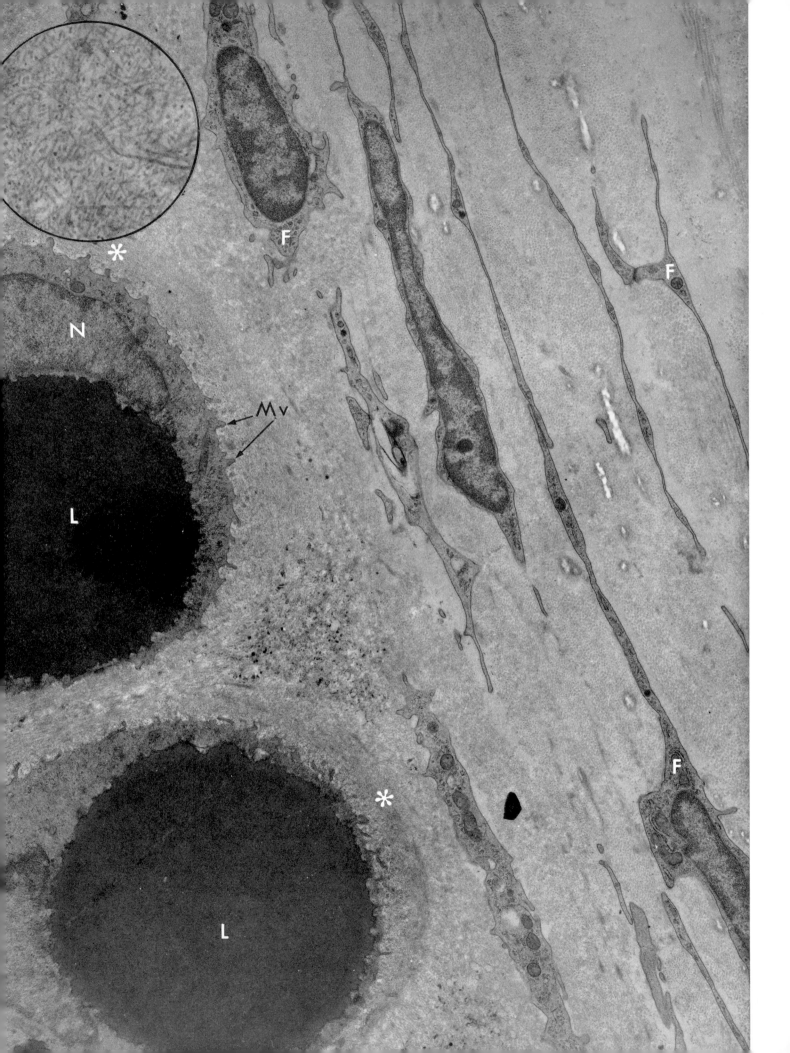

PLATE 29

Cartilage and Perichondrium

CARTILAGE is one of the principal types of vertebrate connective tissues. In some groups, i.e., the elasmobranch fishes, a cartilaginous skeletal system exists throughout life, but in most vertebrates the embryonic skeleton of cartilage is largely replaced by bone. The articulating surfaces of bones, the ends of the ribs, the support of the trachea, nose, and external ear, as well as part of the vertebral column, are examples of cartilaginous structures that persist in adult animals.

Hyaline cartilage, the most common type, is observed in fresh dissections as a translucent material, which though stiff yet retains a certain flexibility and resilience. Superficially it seems to bear little resemblance to connective tissue layers such as the lamina propria examined in Plate 28. Nonetheless, upon microscopical examination, both it and its tough membranous covering, the perichondrium, are easily classified as connective tissues.

The perichondrium, seen at right in this micrograph, comprises lamellae of flattened fibrocytes (F). Their slender processes frequently extend toward their neighbors lying in the same plane to form thin cellular sheets, which are separated by a matrix having abundant collagen fibrils, depicted here primarily in cross section.

Cartilage is not separated from the perichondrium by any cleft or boundary layer. Rather the matrix of one blends into the other with some change in the organization of the collagen fibril bundles. In cartilage, collagen fibrils are thin, and their cross striations are not easily detected (see inset). They are arranged in a three-dimensional lattice that seems to encapsulate (*) the nearly spherical cartilage cells.

Its ground substance distinguishes cartilage from more pliant connective tissues. It is a sulfated mucopolysaccharide linked covalently with a noncollagen protein to make (when hydrated) a stiff gel that lends the cartilage its glassy sheen and stains metachromatically when treated with appropriate dyes. The fibrillar component is not seen in light microscope preparations unless special staining methods are used.

Cartilage cells or chondrocytes differ conspicuously in structure from fibrocytes of the perichondrium and lamina propria (Plate 28). The surfaces of these rounded cells have villuslike projections (Mv) that are intimately associated with the surrounding matrix. The cells therefore fill the lacunae, or spaces, within which they lie. It is not uncommon, as in the present instance, to discover large lipid droplets (L) in the cytoplasm of mature cartilage cells. The large size of these deposits has apparently forced the nucleus (N) to assume a peripheral position and a crescent shape. Indeed the chondrocyte shares a number of cytological features with the adipose cell (cf. Plate 31). In addition, it can be observed that chondrocytes in mature cartilage are separated from one another by a layer of matrix that in places is several micrometers thick. In contrast to bone, no contact between cells exists, and cartilage is not vascularized. Thus, the exchange of metabolites is slower and less efficient than in other tissues. Presumably the lipid stores are designed to compensate for the otherwise limited supply of energy-rich compounds.

The matrix, both fibers and ground substance, is thought to be formed by the chondroblasts, inasmuch as these cells, when isolated in tissue culture, can give rise to normal cartilage tissue and the macromolecules characteristic of the tissue. As already discussed in respect to Plate 28, not all the several mechanisms involved in collagen fiber production are understood, but it is safe to assume that they are essentially the same in all tissues where this protein is produced. On the other hand, the synthesis and secretion of ground substance may be most advantageously studied in cartilage. The high content of sulfur in its polysaccharides provides an opportunity for radioactive labeling. For instance, small pieces of growing cartilage may be excised and incubated with labeled sulfate. A pulse of radioactive material is followed by a "chase" of the same salt in nonradioactive form. The site of the label in the cell at successive times after the pulse is then detected by autoradiographs.

The results of such experiments indicate that the sulfate becomes incorporated first into the mucopolysaccharide in the Golgi region of the cell. Within three minutes after exposure to label, $^{35}SO_4^{--}$ is bound to an insoluble cytoplasmic component that is associated with Golgi membranes and is probably the intracellular precursor of the ground substance. This evidence of Golgi involvement is strengthened by cytochemical studies that localize sulfokinase (an enzyme that catalyzes sulfation of polysaccharides) in the Golgi membranes. The protein moiety is assumed to be synthesized by ribosomes of the rough endoplasmic reticulum and transferred to the Golgi, where it is combined with the polysaccharide.

The Golgi complex of chondroblasts subsequently gives rise to a considerable number of membrane-bounded vesicles that leave the Golgi and empty their contents into the extracellular space. This sequence of events is revealed by the labeled polysaccharide, which leaves the Golgi region, becomes associated with the vesicles, and finally moves into the matrix.

In part, the rough-surfaced ER of chondrocytes in developing systems is probably engaged in making the subunits of collagen. These emerge from the cell by pathways yet to be defined and polymerize into fibrils (inset), which, in association with the ground substance, give cartilage its physical characteristics.

From the tracheal cartilage of the bat
Magnification × 11,000
Inset × 50,000

References

ANDERSON, H. C., and SAJDERA, S. W. The fine structure of bovine nasal cartilage. *J. Cell Biol., 49:*650 (1971).

DZIEWIATKOWSKI, D. D. Intracellular synthesis of chondroitin sulfate. *J. Cell. Biol., 13:*359 (1962).

FEWER, D., THREADGOLD, J., and SHELDON, H. Studies on cartilage. V. Electron microscopic observations on the autoradiographic localization of S^{35} in cells and matrix. *J. Ultrastruct. Res., 11:*166 (1964).

GODMAN, G. C., and LANE, N. On the site of sulfation in the chondrocyte. *J. Cell Biol., 21:*353 (1964).

GODMAN, G. C., and PORTER, K. R. Chondrogenesis studied with the electron microscope. *J. Biophysic. and Biochem. Cytol., 8:*719 (1960).

HAY, E. D. The fine structure of blastema cells and differentiating cartilage cells in regenerating limbs of *Amblystoma* larvae. *J. Biophysic. and Biochem. Cytol., 4:*583 (1958).

MATUKAS, V. J., PANNER, B. J., and ORBISON, J. L. Studies on ultrastructural identification and distribution of protein-polysaccharide in cartilage matrix. *J. Cell Biol., 32:*365 (1967).

MEYER, K. The chemistry of the mesodermal ground substances. *Harvey Lectures, Series 51* (1957) p. 88.

MILLER, E. J., and MATUKAS, V. J. Chick cartilage collagen: a new type of α1 chain not present in bone or skin of the species. *Proc. Nat. Acad. Sci. U.S.A., 64:*1264 (1969).

MINOR, R. R. Somite chondrogenesis. *J. Cell Biol., 56:*27 (1973).

REVEL, J. P. Role of the Golgi apparatus of cartilage cells in the elaboration of matrix glycosamino glycans. *In* Chemistry and Molecular Biology of the Intercellular Matrix. *3.* E. A. Balazs, editor. New York, Academic Press (1970) p. 1485.

———, and HAY, E. D. An autoradiographic and electron microscopic study of collagen synthesis in differentiating cartilage. *Z. Zellforsch., 61:*110 (1963).

ROSENBERG, L., HELLMANN, W., and KLEINSCHMIDT, A. K. Macromolecular models of proteinpolysaccharides from bovine nasal cartilage based on electron microscopic studies. *J. Biol. Chem., 245:*4123 (1970).

SALPETER, M. M. H^3-proline incorporation into cartilage: electron microscopic autoradiographic observations. *J. Morphol., 124:*387 (1968).

SCHUBERT, M. Intercellular macromolecules containing polysaccharide. *Biophys. J., 4* (Suppl.): 119 (1964).

SMITH, J. W. The disposition of proteinpolysaccharide in the epiphyseal plate cartilage of the young rabbit. *J. Cell Sci., 6:*843 (1970).

TRELSTAD, R. L., KANG, A. H., IGARASHI, S., and GROSS, J. Isolation of two distinct collagens from chick cartilage. *Biochemistry, 9:*4993 (1970).

PLATE 30

Osteocytes and Bone

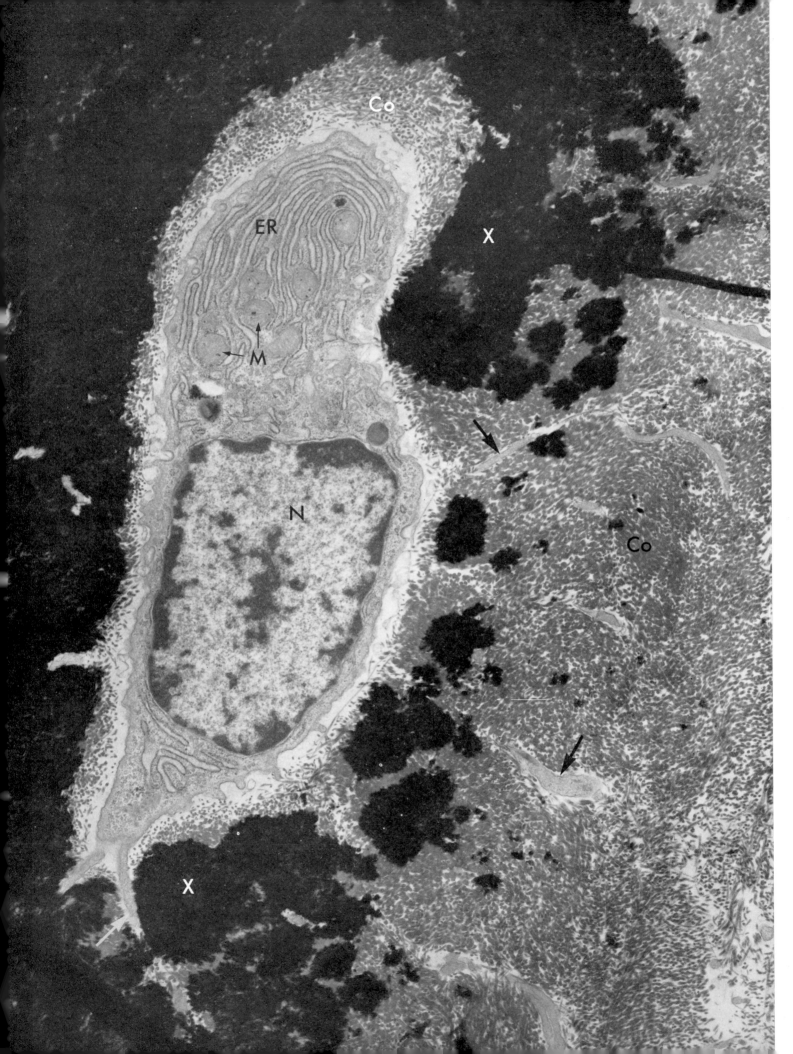

PLATE 30

Osteocytes and Bone

THE osteoblast produces the massive mineralized intercellular matrix of bone. Young cells on the surfaces of developing bone are frequently cuboidal, but as they mature into osteocytes, as shown in this plate, they become stellate. Their thin processes (arrows) extending out from the cell body remain in contact with neighboring cells throughout the life of the tissue. Thus channels filled with living cytoplasm provide pathways by which such cells exchange metabolites. It will be recalled that in cartilage the cells are isolated and that metabolites are transported to and from them by diffusion through the matrix (see Plate 29). Again, in contrast with cartilage, bone is a vascularized tissue, and the cells are never far removed from capillaries.

The intercellular organic matrix of bone consists of a dense feltwork of collagen fibrils (Co) embedded in an amorphous ground substance, a mucopolysaccharide, which differs chemically from that found in cartilage. In bone, in contrast to cartilage, the ground substance constitutes proportionately less of the matrix, and the collagen fibrils make up relatively more. The organic matrix is called the osteoid, and it subsequently achieves great hardness by becoming mineralized. When hydroxyapatite crystals are formed and become densely packed within the matrix, the tissue is called bone. Once the bone cell is encased in the mineralized extracellular material, it is considered a mature osteocyte and cannot contribute further to the growth of the bone unless it is freed from the surrounding matrix as bone is remodeled (see below). New tissue can be added only at the surface where salts are not yet deposited.

The bone cell in this micrograph is only partially encompassed by mineralized material (X), and in its cytological structure there are still indications of its physiological activity. For example, in a part of the cytoplasm seen here at one side of the nucleus (N) there are prominent cisternae of the endoplasmic reticulum (ER). Ribosomes cover the outer surfaces of the membranes, which enclose an amorphous material of medium density. The mitochondria (M) are large and fairly numerous. This appearance of the osteoblast provides indirect evidence that this cell is still engaged in producing proteins for both components of the intercellular matrix.

The synthetic activity of osteoblasts has also been observed more directly by placing them in tissue culture before they have begun to manufacture much, if any, intercellular material. Thus isolated from the organism, they are soon separated from one another by organic matrix. During this period, intracellular polysaccharide-containing granules can be detected, and these presumably contain precursors of the ground substance. Further indication of the direct role of osteoblasts in the formation of bone matrix has come from experiments in which the incorporation of metabolites labeled with radioactive sulfur, carbon, or with tritium have been followed by autoradiography. Thus far the experiments carried out substantiate the concept that the radioactivity is at first concentrated in the cells and only later moves out into the matrix.

The organic matrix of bone, once formed, normally mineralizes, and this phenomenon has been extensively studied. At first, small crystals are found on and within collagen fibrils, and their positions bear a regular relationship to the striations of the fibrils. This observation suggests that the macromolecular structure of collagen governs the initial deposition of hydroxyapatite crystals. The evidence supporting this hypothesis comes from *in vitro* experiments with isolated collagen. For example, it is possible to reconstitute collagen fibrils from extracts of collagenous tissues, both mineralizing and nonmineralizing, and obtain reconstituted fibrils that display a band pattern similar to "native" collagen (i.e., that found *in vivo*) or under other conditions band patterns that are unusual and not found in nature. When these reconstituted collagens are exposed to metastable salt solutions of calcium and phosphorus, only those fibrils with the "native" type banding, regardless of the tissue from which they came, become mineralized. Certain portions of the collagen fibril are therefore said to "seed" the crystals by virtue of their molecular configuration. Recent work indicates that this nucleation occurs at the site of a particular enzyme specifically located with respect to the band pattern along the fibril. This mechanism would require the extracellular fluids to be supersaturated with the appropriate salts, and there is evidence that this is the case in many vertebrates.

127

It should be mentioned in addition that in embryonic osteogenic tissue, where the ground substance is rich in unpolymerized collagen, crystals of hydroxyapatite appear apart from fully formed collagen fibrils.

While hardness tends to become associated in one's mind with permanence, in bone this coupling is invalid. Bone is constantly being remolded: its formation by osteoblasts continues throughout life and is balanced by its resorption by osteoclasts. The latter are a sort of phagocyte. Lying near the free surface of mineralized material, they can in some way break down its structure, destroying both the minerals and the organic matrix.

The harmonious relationship between synthesis and breakdown is subject to influence by many systemic factors. Outstanding among these is the parathyroid hormone, which among others of its effects may stimulate the activity of the osteoclasts. Osteocytes have also proved responsive when exposed to the hormone. When hormonal levels are abnormally high, bone is resorbed, and the blood calcium level rises. This hormone can act directly and in a similar manner on bone tissue isolated in culture. Experiments on isolated mitochondria indicate that the hormone's influence may be due to its ability to affect in a specific way the permeability of mitochondrial membranes in target tissues with respect to calcium and phosphorus and thus to control ionic movements. The importance of this effect in bringing about clinical symptoms of abnormal parathyroid function has not yet been determined.

Vitamin D also plays a role in regulating the mineralization of bone. Its action under normal circumstances, however, is chiefly the indirect one of promoting the absorption of calcium by the intestine.

Other substances too—growth hormone, estrogenic and thyroid hormones—profoundly affect bone structure, but less is known of their mechanisms of action. Vitamin C deficiency apparently prevents the formation of adequate organic matrix, but whether or not it acts directly on osteoblasts has not been discovered. The opportunity to measure qualitatively and quantitatively the action of vitamins and hormones in terms of the extracellular matrix produced invites further investigation of bone.

From the fibula of the mouse
Magnification × 15,500

References

Ascenzi, A., Bonucci, E., and Bocciarelli, D. S. An electron microscope study of osteon calcification. *J. Ultrastruct. Res., 12*:287 (1965).

Decker, J. D. An electron microscope investigation of osteogenesis in the embryonic chick. *Amer. J. Anat., 118*:591 (1966).

Dudley, H. R., and Spiro, D. The fine structure of bone cells. *J. Biophysic. and Biochem. Cytol., 11*:627 (1961).

Fitton-Jackson, S., and Randall, J. T. Fibrogenesis and the formation of matrix in developing bone. *In* Bone Structure and Metabolism, Ciba Foundation Symposium. G. E. W. Wolstenholme and C. M. O'Connor, editors. London, J. and A. Churchill, Ltd. (1956) p. 47.

Glimcher, M. J. The role of the macromolecular aggregation state and reactivity of collagen in calcification. *In* Macromolecular Complexes. M. V. Edds, Jr., editor. New York, The Ronald Press Company (1961) p. 53.

———, and Krane, S. M. The organization and structure of bone, and the mechanism of calcification. *In* Treatise on Collagen 2 (pt B). B. S. Gould, editor. New York, Academic Press (1968) p. 67.

Rasmussen, H. The parathyroid hormone. *Sci. Amer., 204*:56 (April, 1961).

Robinson, R. A., and Cameron, D. A. Bone. *In* Electron Microscopic Anatomy. S. M. Kurtz, editor. New York, Academic Press (1964) p. 315.

Sheldon, H., and Robinson, R. A. Studies on rickets. II. The fine structure of the cellular components of bone in experimental rickets. *Z. Zellforsch. 53*:685 (1961).

Veis, A., Spector, A. R., and Carmichael, D. J. The organization and polymerization of bone and dentin collagens. *Clin. Orthop. Related Res., 66*:188 (1969).

Weinstock, A., and Leblond, C. P. Elaboration of the matrix glycoprotein of enamel by secretory ameloblasts of the rat incisor as revealed by radioautography after galactose-³H injection. *J. Cell Biol., 51*:26 (1971).

———, Weinstock, M., and Leblond, C. P. Autoradiographic detection of ³H-fucose incorporation into glycoprotein by odontoblasts and its deposition at the site of calcification front in dentin. *Calcif. Tissue Res., 8*:181 (1972).

Weinstock, M., and Leblond, C. P. Radioautographic visualization of the deposition of a phosphoprotein at the mineralization front in the dentin of the rat incisor. *J. Cell Biol., 56*:838 (1973).

PLATE 31
Adipose Tissue

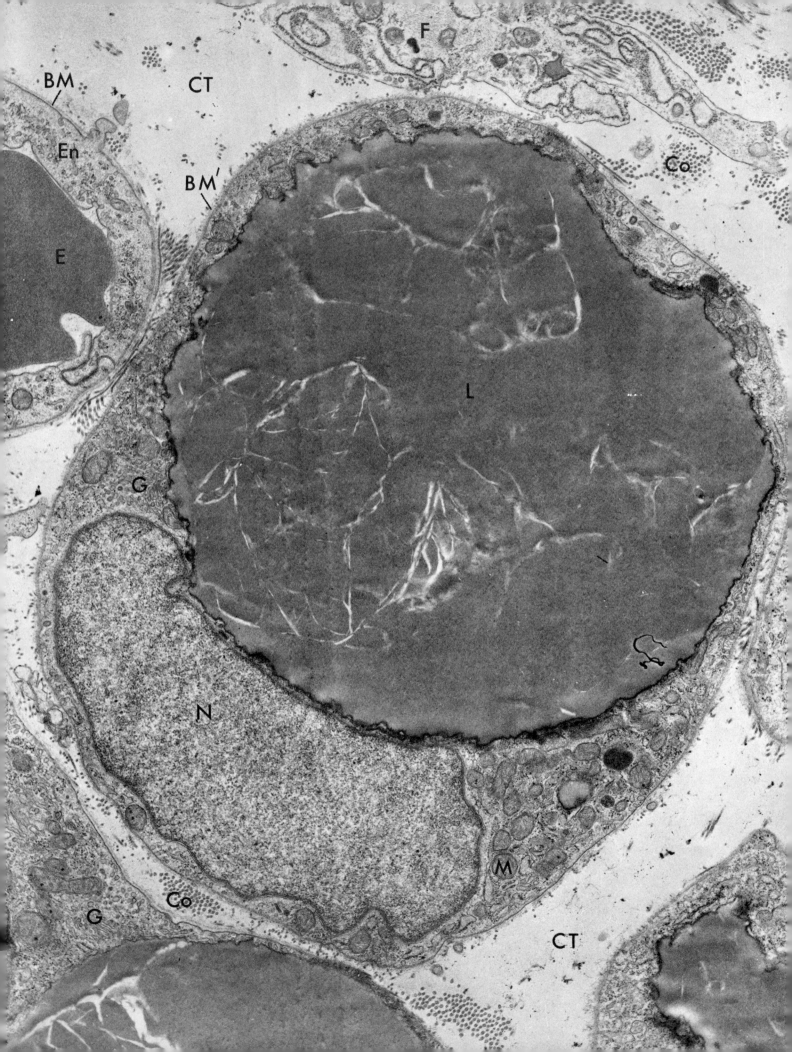

PLATE 31

Adipose Tissue

WHEN caloric intake is excessive, the unused fuel is stored in fatty depots of specialized connective tissue, usually called adipose tissue or white fat. Even the most casual observer is aware that certain subdermal areas of the body are predisposed to accumulate fat, and massive amounts may be stored. In other locations (e.g., scalp, eyelid) fatty deposits do not form even in obese individuals. These considerations, among others, have supported the idea that adipose tissue is a specially differentiated form of connective tissue.

In any event, fat is stored intracellularly in cells that at the outset look very much like fibroblasts. The presumptive fat cell is spindle-shaped with long protoplasmic extensions at each tip. The cytoplasm surrounding the oval nucleus is rich in rough-surfaced endoplasmic reticulum; and mitochondria, as well as Golgi regions, are easily found. When small lipid inclusions appear first at one pole of the cell, it may then be identified as an adipose cell. Lipid droplets next appear at the opposite pole. Small droplets continue to increase in number and coalesce to form larger globules of lipid; at the same time the cell becomes ovoid. In this state, when several lipid droplets are present, the cell is said to be in the multilocular stage. When mature, as shown in this micrograph, and containing only one large fat droplet (L), it is called unilocular. The nucleus (N) is crescent-shaped and has been displaced by the fat toward the periphery of the cell (the so-called signet-ring cell). Golgi membranes (G) may sometimes be seen. Mitochondria (M) are numerous in the perinuclear area. In other regions a thin layer of cytoplasm surrounds the droplet. It is noteworthy that, unlike protein-containing droplets, the lipid is apparently never enclosed by a unit membrane (see Plate 2).

Adipose tissue is highly vascularized, even more so than striated muscle, for numerous capillaries are required to bring absorbed fat to be stored and to carry away fat released when conditions favor depletion. In this example the endothelium (En) of a nearby capillary has an erythrocyte (E) within it.

By starvation and subsequent refeeding it is possible to observe the structural changes that are associated with the depletion and also with the accumulation of fat and to begin to learn in detail the manner in which lipids are transferred from the capillary lumen into the fat droplet. About one hour after a fatty meal, chylomicrons, lipid particles (0.5 to 1.5 μm in diameter) contained in a protein sheath, may be detected in the lymph and blood (cf. Plate 7). Both the liver and adipose tissue play a role in clearing these particles from the circulatory system. In electron micrographs of adipose tissue, chylomicrons may be observed apparently adhering to the luminal surface of the endothelial lining of the vessels, and it is hypothesized that a lipase activity located at that surface frees fatty acids from the triglycerides of the chylomicrons. This interpretation is supported by the fact that no lipid droplets have been detected within the endothelial cells (En), nor have there been discovered in the vessel wall any fenestrae that would permit the passage of intact droplets. Efforts to detect fatty material within the connective tissue have so far gone unrewarded. Therefore the fatty acids are thought to pass through the capillary endothelium (En), its basement membrane (BM), and into the connective tissue ground substance (CT). From the latter, in which collagen fibers (Co) and fibroblasts (F) may be seen, they pass through the basement membrane surrounding the adipose cell (BM') and into its cytoplasm in a form invisible to the microscopist. At the surface of the large droplet the fatty acids are incorporated into neutral fat, the form in which they are stored.

Events involved in the transport of lipid or lipid components across the cytoplasm of the fat cell and into the lipid droplet are less clearly interpreted. In fasted animals that have been refed corn oil the fat cell cytoplasm shows numerous dense osmiophilic particles, which are smaller than ribosomes and tend to aggregate close to fat droplets. In this micrograph they form the dense ring at the periphery of the droplet. These particles, called "lipomicrons," could function in capturing and holding lipids during transport and thus serve to maintain a gradient of concentration that would favor the continuing diffusion of fatty acids and monoglycerides into the cell. The fat, if represented by these granules, would be released into the neutral fat droplet by some reaction occurring at its surface.

During depletion of fat from the storage droplets, the same "lipomicrons" appear in large numbers in the fat cell cytoplasm and are assumed to move the lipid to the cell surface, where it is hydrolyzed for transport to the circulation. An unusual population of small pits in the surfaces of fat cells has also been implicated in the transport of fatty acids through the cytoplasm during fat depletion. These are struc-

turally identical with pits that occur quite generally in endothelial cells and the surfaces of smooth muscle cells (see Plates 10 and 40). Whether their presence is related to the uptake and transport of fatty acids and monoglycerides remains to be demonstrated.

The streaks of low density in the otherwise homogeneous fat droplet (L) are not interpretable except as preparation artifacts.

From the dermis of the newborn rat
Magnification × 7,200

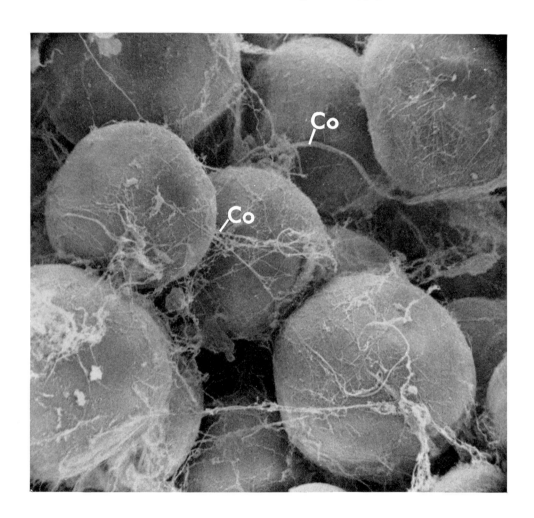

Text Figure 31a

In this cluster of adipose cells, individual units appear as smooth-surfaced spheres when observed by scanning electron microscopy. Delicate strands of connective tissue fibers intertwine the cells and bind them together. These fibers (Co) correspond to those seen in cross and oblique section in Plate 31. The connective tissue ground substance has been removed,

perhaps washed away during preparation of the specimen. The soft, loose nature of the intercellular material is, however, suggested rather well by the SEM image.

From the rat
Magnification × 300

BARRNETT, R. J. The morphology of adipose tissue with particular reference to its histochemistry and ultrastructure. *In* Adipose tissue as an organ. L. W. Kinsell, editor. Springfield, Ill., Charles C Thomas (1962) p. 3.

NAPOLITANO, L. The differentiation of white adipose cells. An electron microscope study. *J. Cell Biol., 18:*663 (1963).

————. The fine structure of adipose tissues. *In* Handbook of Physiology; Adipose Tissue. A. E. Renold and G. F. Cahill, Jr., editors. Washington, D. C., American Physiological Society (1965) p. 109.

SHELDON, H. The fine structure of the fat cell. *In* Fat as a tissue. K. Rodahl and B. Issekutz, Jr., editors. New York, McGraw-Hill (1964) p. 41.

SUTER, E. R., and MAJNO, G. Passage of lipid across vascular endothelium in newborn rats. An electron microscopic study. *J. Cell Biol., 27:*163 (1965).

WASSERMAN, F., and McDONALD, T. F. Electron microscopic study of adipose tissue (fat organs) with special reference to the transport of lipids between blood and fat cells. *Z. Zellforsch., 59:*326 (1963).

WILLIAMSON, J. R. Adipose tissue. Morphological changes associated with lipid mobilization. *J. Cell Biol., 20:*57 (1964).

WOOD, E. N. An ordered complex of filaments surrounding the lipid droplets in developing adipose cells. *Anat. Rec., 157:*437 (1967).

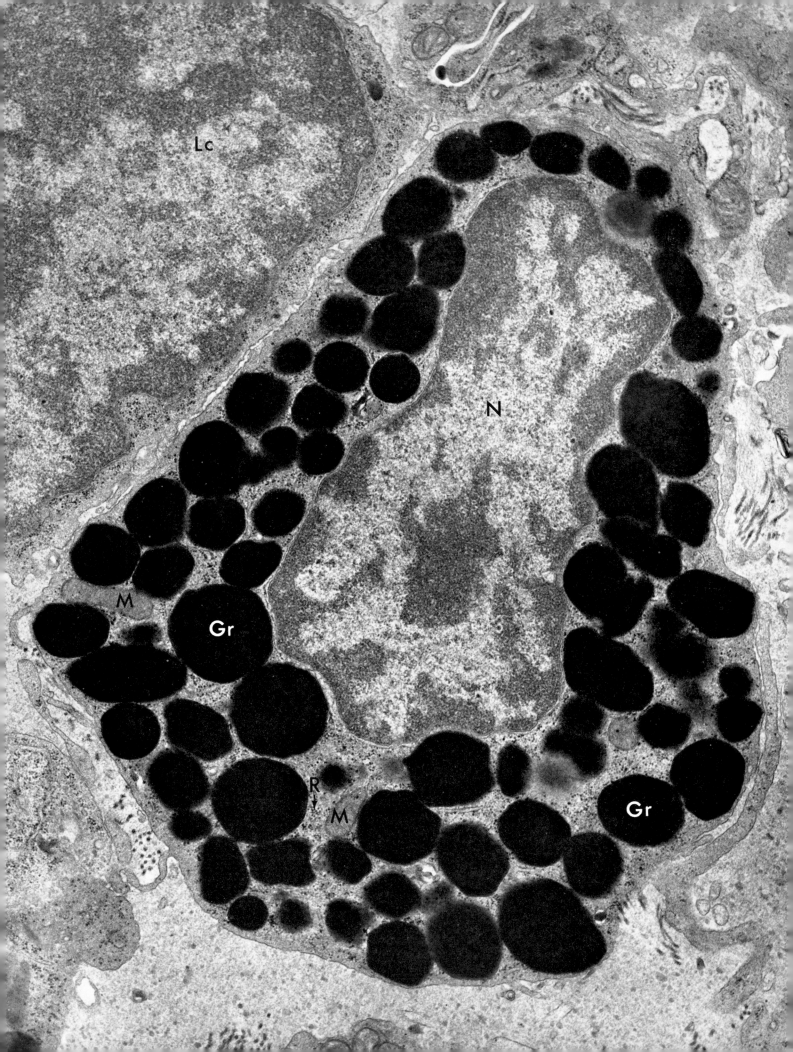

PLATE 32

The Mast Cell

THE mast cells were named about ninety years ago by Ehrlich, who had the impression that their numerous cytoplasmic granules gave them a well-nourished appearance. He designated them as a distinct type of connective tissue cell and noted that their granules stained metachromatically. When toluidine blue is applied to them, the granules assume a purplish-red hue rather than the normal blue color of the dye. This property, now known to be due to the presence of sulfated mucopolysaccharide within the granule, has been used to reveal the widespread distribution of mast cells both in individual animals and in many vertebrate species.

The electron microscope image of the mast cell has confirmed that the numerous granules (Gr) nearly fill the cytoplasm. The mitochondria (M), small and few, were not detected by the light microscope, nor could mitochondrial respiratory enzymes be found in mast cells isolated from fluid of the peritoneal cavity. Free ribosomes (R) may be seen in the cytoplasmic ground substance and along a few profiles of the endoplasmic reticulum. The nucleus (N) displays marginal and central clumping of chromatin material. In brief, these features suggest a fully differentiated cell in which little active synthesis or energy expenditure is occurring. Exposure of mast cells to tritiated thymidine results in only a small amount of labeling, an indication that fully differentiated mast cells are probably not able to divide. The complement of granules is stored, ready to perform its function.

The mast cell depicted here resides in the loose connective tissue of the submucosa. Like other wandering cells it is devoid of a basal lamina. This is especially evident where the mast cell is contiguous with a lymphocyte, shown in part at the upper left (Lc).

Biologists are still uncertain as to the exact function of the mast cell. Since a good deal of information is available concerning the nature of the contents of the granules, speculation on mast cell function centers around the chemical properties of substances localized in them. The main components of mast cell granules are heparin and histamine, both discussed below. In addition, proteases have been found within the granules, and in a few species the vasoconstrictor serotonin (5-hydroxytryptamine) is also present within them.

The mast cell was first suspected of being the source of heparin, a powerful anticoagulant isolated first from the livers of dogs, when it was noted that heparin content and mast cell number exhibited a positive correlation in a number of tissues. Isolation of mast cells from the peritoneal cavity and the discovery of mast cell tumors provided opportunities for testing directly and thereby confirming that heparin, a sulfated mucopolysaccharide, is indeed contained in the mast cell granules.

The normal function of heparin is less clear than its origins. Its pharmacological effect, that is, its action as an anticoagulant administered clinically in large doses, may not, and probably does not, correspond to its physiological function. Indeed, its anticoagulating ability depends upon its source, and heparin from certain species of animals may show little of this property. In the mast cell granule, heparin seems to be bound to a structural protein, and this association survives during normal functioning of the cell (see below).

Mast cell granules also contain histamine (2-[4-imidazolyl] ethylamine), although this compound is not localized in them exclusively. Release of histamine leads to increased permeability of the capillaries with consequent flow of plasma proteins into the intercellular space. The resulting edema is characteristic of allergic reactions, including anaphylactic shock. As a further consequence of histamine release, leukocytes invade the edematous zone, and phagocytosis is stimulated.

In order to determine the fate of mast cell granules the histological effects of substances known to release histamine from tissues have been studied. One of these releasers—compound 48/80—first reduces the density of mast cell granules. A halolike space appears, separating the granule from the membrane that encloses it. The staining properties of granules altered in this manner indicate that heparin is still present within them. It has been proposed, therefore, that the heparin with its high electronegative charge may function in holding histamine within the granule. At any rate, histamine release does occur without complete disruption of mast cell granules, although with high enough doses of histamine

135

releaser the granules may be ejected from the cell, free of their membranous coverings.

The origin of the mast cell granules relative to the cell's fine structure holds some interest, for it seemingly correlates with the known presence of heparin, histamine, and protein in the mature granules. Early in the morphogenesis of these cells, small dense granules appear in vesicles of the Golgi zone. Later these appear to combine with larger vesicles of ER origin suspected of containing protein. Such observations are consistent with the accepted view of the Golgi as the site of mucopolysaccharide formation (heparin) and the ER as the source of protein (cf. Plates 12, 28, and 29).

From the intestinal submucosa of the rat
Magnification × 27,400

References

BENDITT, E. P., HOLCENBERG, J., and LAGUNOFF, D. The role of serotonin (5-hydroxytryptamine) in mast cells. *Ann. N. Y. Acad. Sci., 103* (Article 1):179 (1963).

————, and LAGUNOFF, D. The mast cell: Its structure and function. *Progr. Allergy, 8:*195 (1964).

BLOOM, G. D., and HAEGERMARK, O. A study on morphological changes and histamine release induced by compound 48/80 in rat peritoneal mast cells. *Exp. Cell Res., 40:*637 (1965).

COMBS, J. W. Maturation of rat mast cells. An electron microscope study. *J. Cell Biol., 31:*563 (1966).

LAGUNOFF, D. Contributions of electron microscopy to the study of mast cells. *J. Invest. Dermatol., 58:*296 (1972).

————. The mechanism of histamine release from mast cells. *Biochem. Pharmacol., 21:*1889 (1972).

————. Membrane fusion during mast cell secretion. *J. Cell Biol., 57:*252 (1973).

————, PHILLIPS, M., ISERI, O. A., and BENDITT, E. P. Isolation and preliminary characterization of rat mast cell granules. *Lab. Invest., 13:*1331 (1964).

RILEY, J. F. Functional significance of histamine and heparin in tissue mast cells. *Ann. N. Y. Acad. Sci., 103* (Article 1):151 (1963).

RÖHLICH, P., ANDERSON, P., and UVNÄS, B. Electron microscope observations on compound 48/80-induced degranulation in rat mast cells. *J. Cell Biol., 51:*465 (1971).

SCHILLER, S. Mucopolysaccharides of normal mast cells. *Ann. N. Y. Acad. Sci., 103* (Article 1):199 (1963).

SMITH, D. E. Electron microscopy of normal mast cells under various experimental conditions. *Ann. N. Y. Acad. Sci., 103* (Article 1):40 (1963).

THIÉRY, J. P. Etude au microscope électronique de la maturation et de l'excrétion des granules des mastocytes. *J. Microscopie, 2:*549 (1963).

PLATE 33

The Erythroblast and Erythrocyte

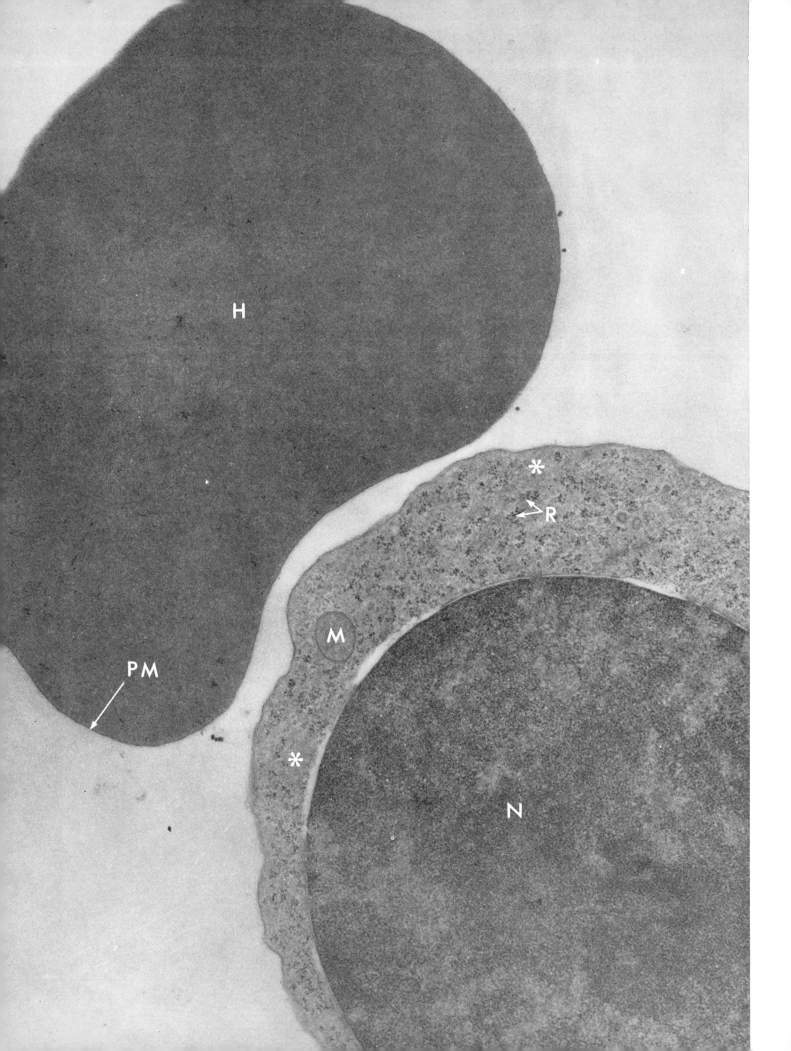

PLATE 33

The Erythroblast and Erythrocyte

THE mammalian erythrocyte or red blood corpuscle is a striking example of a specialized cell. The differentiated form, shown in the upper left-hand portion of this figure, develops in hematopoietic tissue. The limiting membrane (PM), seen as a line, encloses a mass of material (H) consisting mainly of hemoglobin. This respiratory pigment is notable in electron micrographs for its denseness, which is largely due to its iron content. The nucleus has already been extruded from this red cell, and cytoplasmic organelles have disintegrated so that the differentiated unit contains very little except the characteristic respiratory pigment. In the image shown here, a few particles, probably ribosomes, persist, indicating that protein synthesis may still be continuing. In completely mature forms circulating in the blood stream, even these particles disappear, and all the synthetic machinery of the cell is closed down.

An earlier stage in this differentiation is shown by the erythroblast present at the lower right. This cell, which is probably a polychromatophilic erythroblast and which is common in hematopoietic tissue, has ceased to divide; but unlike the mature erythrocyte, it still possesses a nucleus (N) with nuclear envelope, a few mitochondria (M), and small groups of ribosomes (R) separated from each other by a cytoplasmic ground substance or stroma already accumulating hemoglobin (*). Such groups of ribosomes, referred to as polysomes, represent the assembling unit in protein synthesis. It is noteworthy that in the differentiation of this cell the production of protein for intracellular storage does not involve the membranous systems; that is, the endoplasmic reticulum and the Golgi complex, normally present in the cytoplasm of most cells (see Plates 2 and 11). Free or unattached ribosomes are commonly found in embryonic cells and are uniformly associated with the processes of cellular growth and differentiation in which the products of synthesis are retained.

During the course of red cell differentiation, as already indicated above, ribosomes and elements of the ER gradually disappear from the cytoplasm as hemoglobin increases in prominence. Disappearance of ribosomes is correlated with termination of their role in hemoglobin synthesis. Simultaneously, and possibly in relation to these events, the nucleoli fragment and essentially fade from view. Marked clumping of heterochromatic material also occurs as differentiation progresses, and this change is assumed to reflect a cessation in chromosomal production of messenger RNA (cf. Plate 4). Nuclear pores, which normally open into zones of the nucleoplasm between masses of heterochromatin, become less numerous as the cell matures.

Thus micrographs of differentiating erythroblasts provide several signs consistent with the general atrophy of nuclear activity. In mammalian erythroblasts extrusion of the nucleus follows, occurring as one of the last steps in differentiation. First the erythroblast becomes highly polarized as the future cytoplasm of the red cell and the nucleus move into separate parts of the cell. Then vesicles assemble near the neck or isthmus connecting the two parts. Next the vesicles coalesce to form cisternal structures that unite with the plasma membrane of the cell so as to separate the nuclear-containing portion from the young red cell. This sequence of events is reminiscent of those shown by the megakaryocyte in the formation of blood platelets (see Plate 35). It is to be noted that the extruded red cell nucleus retains a thin layer of cytoplasm, including mitochondria, and remains enclosed by the plasma membrane. The potentialities of this attenuated cell have not been tested, so that it is not known whether it may continue to form additional erythrocytes. Probably its usual fate is to be engulfed and broken down by a reticuloendothelial cell, a kind of phagocyte found in hematopoietic tissue (cf. Plate 37).

Sinusoids are characteristic of hematopoietic tissue. These are wide bore capillaries with thin endothelial linings. Usually gaps are observed between endothelial cells. After completing their maturation, red cells are able to slip through these interruptions in the vascular lining and join circulating blood cells.

From the liver of a rat embryo
Magnification $\times$ 30,000

139

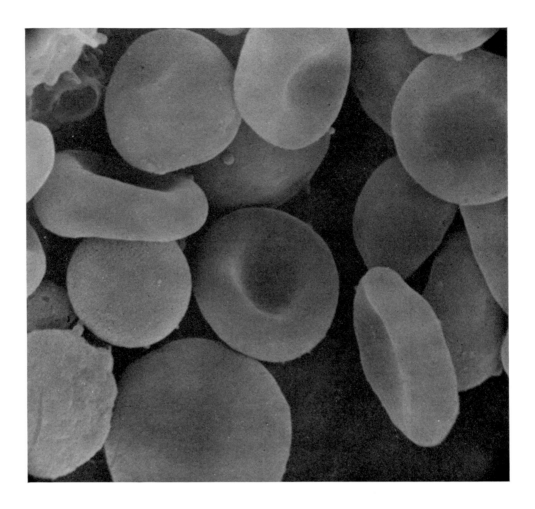

Text Figure 33a

This scanning electron micrograph of red blood cells graphically confirms that these cells are biconcave disks. The red cells are sensitive osmometers and can be easily distorted by suspending them in a medium not exactly isotonic to blood plasma. The swelling shown by some of the cells in this field is probably due to movement of water into them from the medium used during preparation.

Red cell morphology, although familiar, is extraordinary. The shape of the cells is constructed so as to allow maximum surface area for gaseous exchanges between hemoglobin and the extracellular environment. The outer surface is exceptionally smooth. The limiting membrane has a degree of flexibility, allowing its deformation as cells are squeezed through spaces between sinusoidal endothelial cells or into narrow capillaries. After such distortions the red cell can return to its normal shape; presumably the organization of its internal proteins and their relationship to the plasma membrane are somewhat elastic but still rigid enough to recoil once abnormal pressures are no longer present.

From the rat
Magnification × 7,500

References

BERMAN, I. The ultrastructure of erythroblastic islands and reticular cells in mouse bone marrow. *J. Ultrastruct. Res., 17:*291 (1967).

CAMPBELL, G. LeM., WEINTRAUB, H., MAYALL, B. H., and HOLTZER, H. Primitive erythropoiesis in early chick embryogenesis. II. Correlation between hemoglobin synthesis and the mitotic history. *J. Cell Biol.. 50:*669 (1971).

FANTONI, A., DE LA CHAPELLE, A., RIFKIND, R. A., and MARKS, P. A. Erythroid cell development in fetal mice: Synthetic capacity for different proteins. *J. Mol. Biol., 33:*79 (1968).

GRASSO, J. A., SWIFT, H., and ACKERMAN, G. A. Observations on the development of erythrocytes in mammalian fetal liver. *J. Cell Biol., 14:*235 (1962).

KOVACH, J. S., MARKS, P. A., RUSSELL, E. S., and EPLER, H. Erythroid cell development in fetal mice: ultrastructural characteristics and hemoglobin synthesis. *J. Mol. Biol., 25:*131 (1967).

ORLIC, D., GORDON, A. S., and RHODIN, J. A. G. An ultrastructural study of erythropoietin-induced red cell formation in mouse spleen. *J. Ultrastruct. Res., 13:*516 (1965).

PRESTON, F. E., and SHAHANI, R. T. Surface ultramicroscopy of neonatal erythrocytes. *The Lancet, 1:*1177 (1970).

RIFKIND, R. A., CHUI, E., and EPLER, H. An ultrastructural study of early morphogenetic events during the establishment of fetal hepatic erythropoiesis. *J. Cell Biol., 40:*343 (1969).

SIMPSON, C. F., and KLING, J. M. The mechanism of denucleation in circulating erythroblasts. *J. Cell Biol., 35:*237 (1967).

WARNER, J. R., RICH, A., and HALL, C. E. Electron microscope studies of ribosomal clusters synthesizing hemoglobin. *Science, 138:*1399 (1962).

WEISS, L. The structure of bone marrow. Functional interrelationships of vascular and hematopoietic compartments in experimental hemolytic anemia. *J. Morph., 117:*467 (1965).

WICKRAMASINGHE, S. N., COOPER, E. H., and CHALMERS, D. G. A study of erythropoiesis by combined morphologic, quantitative cytochemical and autoradiographic methods. Normal human bone marrow, vitamin B_{12} deficiency and iron deficiency anemia. *Blood, 31:*304 (1968).

ZAMBONI, L. Electron microscopic studies of blood embryogenesis in humans. II. The hemopoietic activity in in the fetal liver. *J. Ultrastruct. Res., 12:*525 (1965).

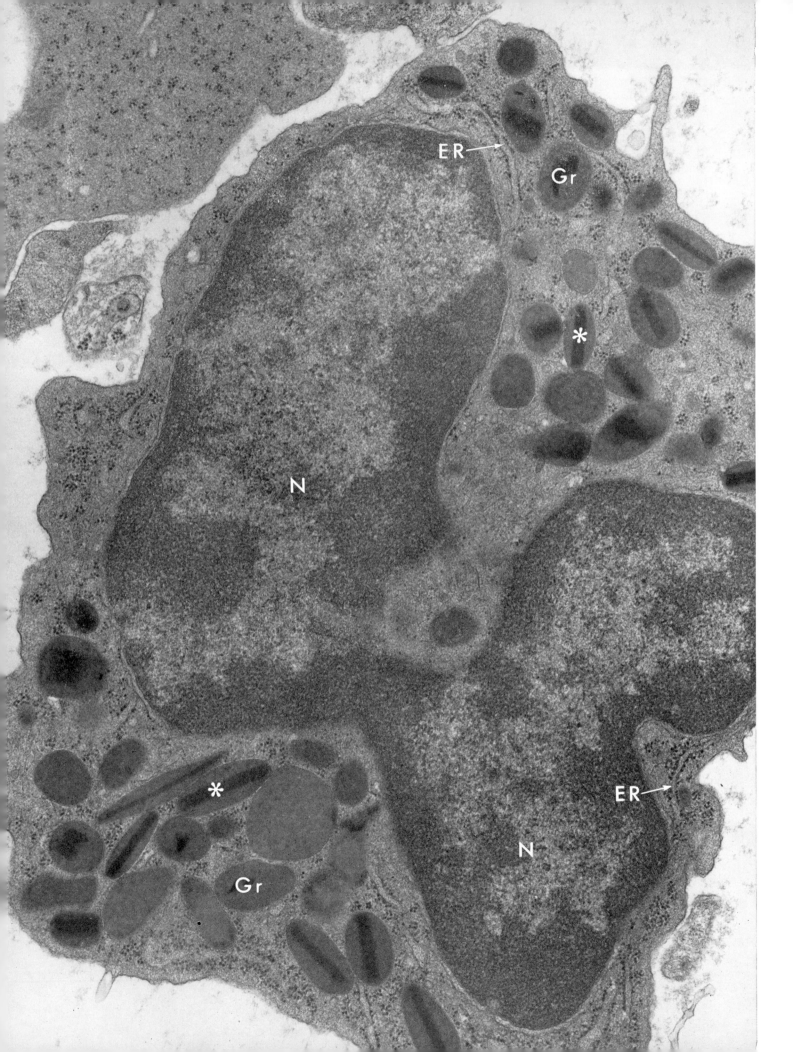

PLATE 34

The Eosinophilic Leukocyte

THE eosinophils with their many cytoplasmic granules have long attracted the eye of the microscopist, but only recently has some progress been made in understanding their function. This electron micrograph shows the structure of an eosinophil, which although still in the bone marrow is in the final stages of its development. This section includes a large part of the prominent bilobed nucleus (N). In the peripheral cytoplasm there are several cisternae of the endoplasmic reticulum (ER), which enclose an amorphous material of low density and support numerous ribosomes on their surfaces. These leukocytes have a large Golgi complex close to the nucleus, but in this section it happens not to be included.

The outstanding components of the cell are the numerous granules (Gr) characteristic of this leukocyte. These are biconvex disks surrounded by a thin membrane and composed of a matrix material of medium density. In favorable sections one or more rods of dense, crystalline material (*) lie in the equatorial region of the granules. Recently the structure of the crystal has been identified as a cubic lattice, but as yet its composition has not been determined. A peroxidase peculiar to eosinophils, and the lysosomal enzyme, acid phosphatase, have been localized in the matrix of the granule. The matrix also contains an array of other enzymes typical of lysosomes found in various other cells, including neutrophils. However, unlike granules of neutrophils, they seem to lack specific bacteriocidal agents.

The formation of the specific granules of eosinophils closely resembles that of other secretory granules. Early forms appear as condensation vacuoles in the Golgi zone and develop into mature granules as materials are added to them from both the ER and Golgi sources.

Eosinophils are found in considerable numbers in skin, as well as in the tissues of the respiratory and digestive tracts. They are therefore in a suitable position to play a role in defense of the organism against various pathogenic agents. It has been shown that under certain conditions eosinophils become phagocytic and that granules adjacent to the ingested material burst. The hydrolytic enzymes within the granules are thought to enter the phagocytic vacuoles, there to break down the ingested material. This chain of events, as well as the attraction of eosinophils to specific locations in tissues, is apparently stimulated by antigen-antibody reactions.

From the bone marrow of the rat
Magnification × 37,000

References

ARCHER, G. T. The function of the eosinophil. *Bibl. Haematol.*, 29:71 (1968).
———, AIR, G., JACKAS, M., and MORELL, D. B. Studies on rat eosinophil peroxidase. *Biochim. Biophys. Acta*, 99:96 (1965).
———, and HIRSCH, J. G. Isolation of granules from eosinophil leucocytes and study of their enzyme content. *J. Exp. Med.*, 118:227 (1963).
———, and HIRSCH, J. G. Motion picture studies on degranulation of horse eosinophils during phagocytosis. *J. Exp. Med.*, 118:287 (1963).
BAINTON, D. F., and FARQUHAR, M. G. Segregation and packaging of granule enzymes in eosinophilic leukocytes. *J. Cell Biol.*, 45:54 (1970).
COTRAN, R. S., and LITT, M. The entry of granule-associated peroxidase into phagocytic vacuoles of eosinophils. *J. Exp. Med.*, 129:1291 (1969).
DUNN, W. B., HARDIN, J. H., and SPICER, S. S. Ultrastructural localization of myeloperoxidase in human neutrophil and rabbit heterophil and eosinophil leukocytes. *Blood*, 32:395 (1968).

FEDORKO, M. E. Formation of cytoplasmic granules in human eosinophilic myelocytes: an electron microscope autoradiographic study. *Blood*, 31:188 (1968).
HARDIN, J. H., and SPICER, S. S. An ultrastructural study of human eosinophil granules: maturational stages and pyroantimonate reactive cation. *Amer. J. Anat.*, 128:283 (1970).
———, and SPICER, S. S. Ultrastructural localization of dialyzed iron-reactive mucosubstance in rabbit heterophils, basophils, and ecsinophils. *J. Cell Biol.*, 48:368 (1971).
MILLER, F., deHARVEN, E., and PALADE, G. E. The structure of eosinophil leucocyte granules in rodents and in man. *J. Cell Biol.*, 31:349 (1966).
SEEMAN, P. M., and PALADE, G. E. Acid phosphatase localization in rabbit eosinophils. *J. Cell Biol.*, 34:745 (1967).
WETZEL, B. K., HORN, R. G., and SPICER, S. S. Fine structural studies on the development of heterophil, eosinophil, and basophil granulocytes in rabbit. *Lab. Invest.*, 16:349 (1967).

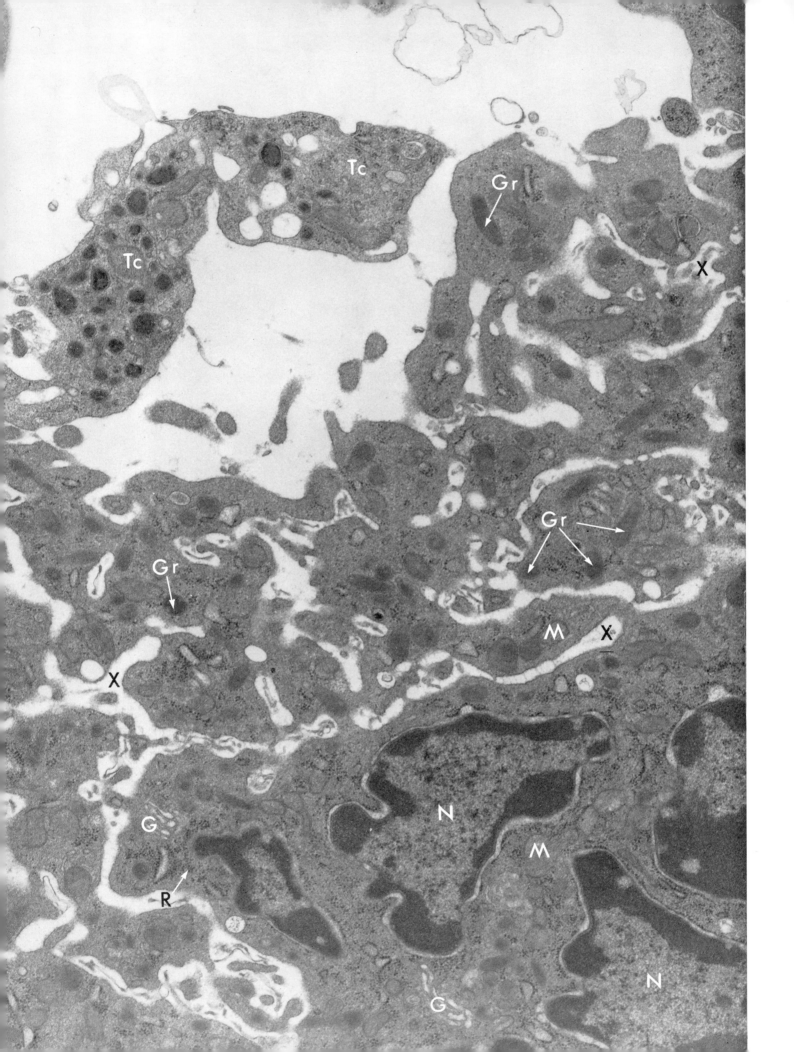

PLATE 35

The Megakaryocyte

THE megakaryocyte is easily identified in the blood-forming tissue of mammals because of its large multilobular nucleus, from which its name is derived, and its extensive cytoplasm, which makes it a giant among myeloid elements. Early in this century careful examinations by light microscopy indicated that this cell gives rise to the blood platelets or thrombocytes, which take part in the formation of blood clots. More recently further information on the manner of their production has come from studies with the electron microscope.

Only isolated portions of the nucleus (N) are evident in this micrograph of a megakaryocyte, and the cytoplasm, though abundant, is only partially included. The latter especially possesses unusual structural features. In addition to mitochondria (M), Golgi regions (G), and ribsomes (R) the cytoplasm is dotted with spherical-to-oblong granules of moderate density (Gr), which are apparently membrane limited. Evidence from electron micrographs suggests that these granules are formed in the Golgi region.

Individual platelets are, however, small cells devoid of nuclei. They contain within a plasma membrane such cytoplasmic components as mitochondria, ribosomes, and vesicles of the endoplasmic reticulum in addition to the characteristic granules just mentioned. Two such cells are seen at Tc, upper left.

Platelets are believed to form from the megakaryocyte by a process in which the cytoplasm of the parent cell is first partitioned and then fragmented along the partitions. Initially in this process curtains of "platelet demarcation vesicles" form. In thin sections these appear as strings of membrane-bounded elements. Later the vesicles coalesce, and become tubular; still later they form large flattened cisternae, the "platelet demarcation membranes" (X). Thus a system of vesicular septa develops, and each separated area is eventually completely detached from the megakaryocyte (see upper left in micrograph). This is essentially a process of secretion similar to that involved in red cell formation (see Plate 33).

The partitions or "platelet demarcation membranes" obviously become the limiting membrane of the platelets after they are freed from the megakaryocyte. These membranes hold more than the usual interest, for they seem to contain a factor (thromboplastin) that activates prothrombin and thus sets the stage for fibrin clot formation. The tendency of platelets to clump is due to a layer of fibrinogen adsorbed on their surfaces.

The dense granules within the platelets (Gr) are membrane limited and slightly smaller than mitochondria. When isolated by the techniques of cell fractionation, they have been shown to be rich in acid phosphatase and β-glucuronidase, facts which probably indicate their similarity to lysosomal granules of other cells (see Plate 15). Though the granules are released during clot formation and appear to disintegrate at that time, their function has not been established. Other compounds, such as serotonin and histamine, are also released from platelets during clot formation, but their localization in the cells is not entirely certain. "Very dense" granules, which are sometimes seen, are believed to contain serotonin. For some unexplained reason platelets not involved in clotting may be active in the phagocytosis of foreign bodies such as viruses.

From the bone marrow of the rat
Magnification $\times$ 30,000

BEHNKE, O. Electron microscopic observations on the membrane systems of the rat blood platelet. *Anat. Rec., 158:*121 (1967).

―――. An electron microscope study of the rat megacaryocyte. II. Some aspects of platelet release and microtubules. *J. Ultrastruct. Res., 26:*111 (1969).

―――, and ZELANDER, T. Filamentous substructure of microtubules of the marginal bundle of mammalian blood platelets. *J. Ultrastruct. Res., 19:*147 (1967).

CRONKITE, E. P., BOND, V. P., FLIEDNER, T. M., PAGLIA, D. A., and ADAMIK, E. R. Studies on the origin, production and destruction of platelets. *In* Blood Platelets. S. A. Johnson, R. Monto, J. Rebuck, and R. C. Horn, editors. Boston, Little, Brown and Company (1961) p. 595.

HAN, S. S., and BAKER, B. L. The ultrastructure of megakaryocytes and blood platelets in the rat spleen. *Anat. Rec., 149:*251 (1964).

JØRGENSEN, L., ROWSELL, H. C., HOVIG, T., and MUSTARD, J. F. Resolution and organization of platelet-rich mural thrombi in carotid arteries of swine. *Amer. J. Pathol., 51:*681 (1967).

―――, ROWSELL, H. C., HOVIG, T., GLYNN, M. F., and MUSTARD, J. F. Adenosine-diphosphate-induced platelet aggregation and myocardial infarcts in swine. *Lab. Invest., 17:*616 (1967).

MARCUS, A. J., and ZUCKER-FRANKLIN, D. Enzyme and coagulation activity of subcellular platelet fractions. *J. Clin. Invest., 43:*1241 (Abstract) (1964).

MOVAT, H. Z., MUSTARD, J. F., TAICHMAN, N. S., and URIUHARA, T. Platelet aggregation and release of ADP, serotonin and histamine associated with phagocytosis of antigen-antibody complexes. *Proc. Soc. Exp. Biol. Med., 120:*232 (1965).

MOVAT, H. Z., WEISER, W. J., GLYNN, M. F., and MUSTARD, J. F. Platelet phagocytosis and aggregation. *J. Cell Biol., 27:*531 (1965).

NATHANIEL, E. J. H., and CHANDLER, A. B. Electron microscopic study of adenosine diphosphate-induced platelet thrombi in the rat. *J. Ultrastruct. Res., 22:* 348 (1968).

RODMAN, N. F., JR., PAINTER, J. C., and MCDEVITT, N. B. Platelet disintegration during clotting. *J. Cell Biol., 16:* 225 (1963).

SCHMID, H. J., JACKSON, D. P., and CONLEY, C. L. Mechanism of action of thrombin on platelets. *J. Clin. Invest., 41:*543 (1962).

SILVER, M. D., and GARDNER, H. A. The very dense granule in rabbit platelets. *J. Ultrastruct. Res., 23:*366 (1968).

THIÉRY, J. P., and BESSIS, M. Mécanisme de la plaquetto-génèse: étude *in vitro* par la microcinématographie. *Rev. hémat., 11:*162 (1956).

TRANZER, J. P., DA PRADA, M., and PLETSCHER, A. Storage of 5-hydroxytryptamine in megakaryocytes. *J. Cell Biol., 52:*191 (1972).

VASSALLI, P., SIMON, G., and ROUILLER, C. Ultrastructural study of platelet changes initiated *in vivo* by thrombin. *J. Ultrastruct. Res., 11:*374 (1964).

YAMADA, E. The fine structure of the megakaryocyte in the mouse spleen. *Acta Anat., 29:*267 (1957).

ZUCKER-FRANKLIN, D., NACHMAN, R. L., and MARCUS, A. J. Ultrastructure of thrombosthenin, the contractile protein of human blood platelets. *Science, 157:*945 (1967).

PLATE 36
The Thymus

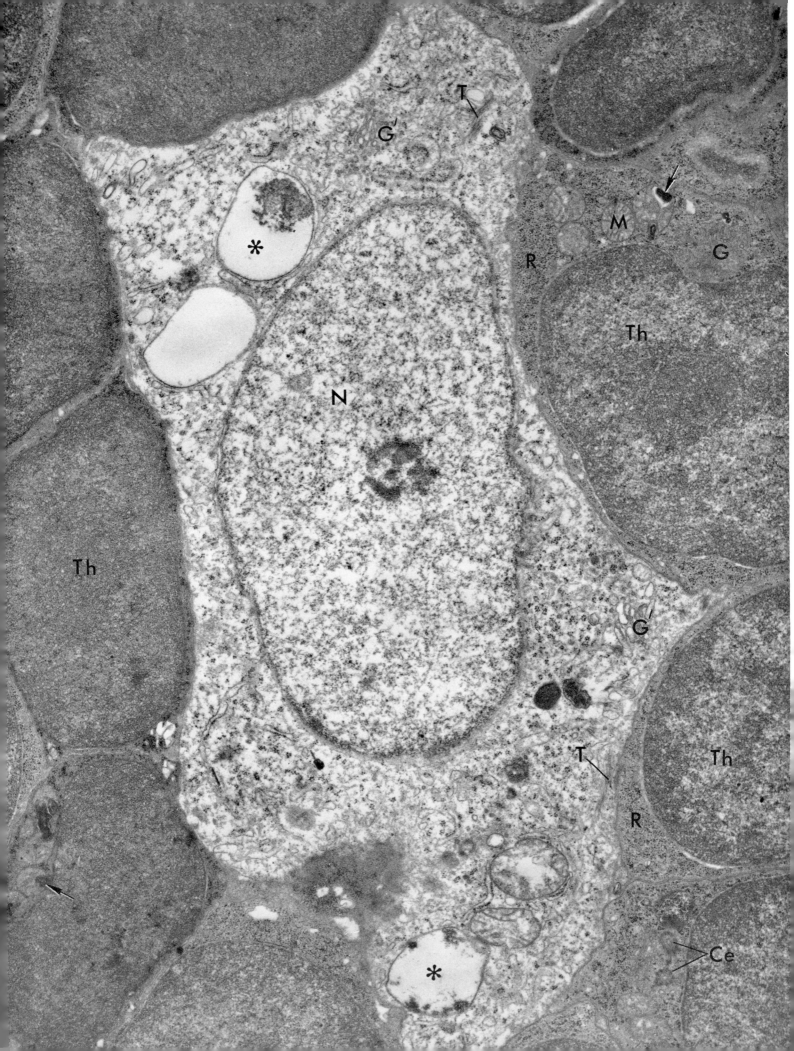

PLATE 36

The Thymus

THE thymus, an organ that is well developed in young animals but that decreases greatly in size at the time the animal becomes sexually mature, has only recently been divested of some of the mystery regarding its function. Noticing its dense aggregations of cells resembling lymphocytes, earlier investigators believed it might be an important source of these cells. Now, however, it is known to exert control over the development of immunological defense mechanisms that lead to rejection of antigenic molecules as well as cells and tissues not derived from the animal itself.

Removal of the thymus from mice on the day of their birth has remarkable effects on the health and life of the animal. Once the technical skill for this delicate operation was acquired, it was shown that the resulting "wasting disease" stemmed from inability of the animal to make antibody and to reject foreign cells and tissues. Invasion by foreign organisms led to the early death of the defenseless animals.

Thymectomized animals never develop plasma cells, which produce antibody (see Plate 1), and they lack circulating lymphocytes as well as aggregations of lymphocytes characteristic of normal lymph nodes and spleen. The lymphocytes are probably responsible for the phenomenon of "delayed hypersensitivity." A good example of this is the inevitable rejection of tissue grafts derived from all other individuals except from an identical twin. The fetal thymus is therefore said to "seed" the tissues where immunologically reactive cells are produced.

In the thymus prior to involution the most numerous cell type is the thymocyte (Th). These cells closely resemble lymphocytes in that the chromatin in their nuclei is densely clumped. The cytoplasm is rich in ribosomes (R), and its volume is small relative to that of the nucleus. Golgi regions (G), centrioles (Ce), and mitochondria (M) may be aggregated at one pole of the nucleus. Small myelin figures (arrows) are frequently enclosed within the mitochondria.

Initially, thymocytes are produced in the thymus at a high rate. Cells that are destined to proliferate enlarge before dividing, and a series of mitotic divisions follows, leading to the production of the small thymocytes. After the seeding of other lymphoid tissues, lymphocytes no longer leave the thymus in appreciable numbers. It seems that the lymphoid organs, especially the spleen and lymph nodes, once having acquired lymphocytes, are immunologically competent.

Interest in the thymus was greatly stimulated by the initial discovery of its importance, and the ensuing investigations revealed an additional function; namely, that it serves as an endocrine organ. As evidence of this, it has been found that transplants of thymus tissue are adequate replacement for glands removed surgically soon after birth. This holds true even when the thymus transplant is enclosed in a porous chamber that allows exchange of body fluids but prevents escape of thymus cells into the host. The convincing test for hormonal involvement is that animals with such transplants respond normally when their immunological defense mechanisms are challenged. Therefore one can conclude that the mammalian thymus behaves as if it were secreting a hormone or hormones that stimulate plasma cells and lymphocytes to proliferate and to form antibody when antigens are introduced.

The thymus hormone is now being isolated and characterized, and examination of the thymus does disclose the existence of cells that may be secretory. One such cell is seen in this micrograph (N). Its nuclear content is thinly dispersed, and there is notably little clumping of chromatinic material. The cytoplasm, also of low density, extends into long processes that attach to other cells of the same nature by means of desmosomes to form a reticular framework within which the lymphocytes are held. Bundles of tonofilaments (T) lie in the cytoplasm. Because of these features and also because they may possess cilia and have one surface resting on a basement membrane, these cells are considered to be epithelial. Vesicular (∗) and granular inclusions of the epithelial cells have been examined in regard to their staining properties and their ability to incorporate isotopically labeled metabolites. Such studies have demonstrated that these structures contain sulfated acid mucopolysaccharides, and cytological observations suggest that the vesicles are formed within the Golgi regions (G′) and may contain the secretory product of these cells, that is, the thymus hormone.

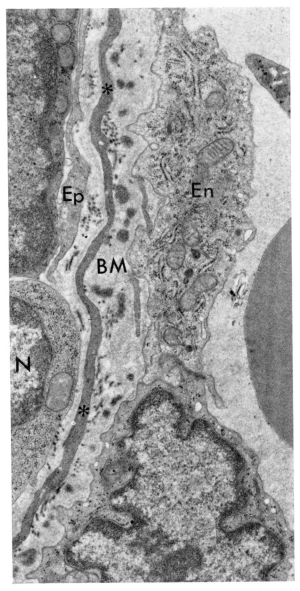

Text Figure 36a

The capillary endothelium (En) of thymus vessels is relatively thick (compare Plates 10 and 24) and lacks obvious fenestrae. An extensive layer of basement membrane material (BM) supports the endothelial cells and encompasses pericytes (*). Processes of thymus epithelial cells (Ep) and lymphocytes (nucleus, N) border the capillary wall.

From the thymus of the newborn rat
Magnification × 9000

While the thymus stimulates lymphopoiesis, it does not itself normally produce any antibody-forming cells. Indeed it lacks germinal centers, present in lymph nodes and the white pulp of the spleen, in which immunologically competent cells are produced in response to suitable stimuli. The lack of response by the thymus may perhaps be related to the fact that little stimulating antigenic material can traverse the "blood-thymus barrier." This barrier in the newborn rats has been found to be only partially effective, whereas in adults it is well established.

Thus it would appear that the capillaries of the thymus are the sites at which substances entering the gland may be monitored. In text figure 36a, included to illustrate this barrier, the capillary endothelium (En) proves to be one of considerable thickness when compared to many other endothelia (cf. Plates 10 and 25). Furthermore it lacks fenestrae and is supported by a thickened layer of basement membrane material (BM), which encloses collagen fibrils. Pericytes are surrounded by this material and may form an almost complete cellular wrapping around the vessel. Processes of epithelial cells (Ep) often extend along the connective tissue of the capillary wall, but lymphocytes (nucleus, N) may abut on it as well. It seems likely that the unusual and rather elaborate structure of this blood vessel and supporting tissues constitutes the physical basis of the "blood-thymus barrier."

Germinal centers appear in the thymus in a certain few of the autoimmune diseases (e.g., hemolytic anemia and myasthenia gravis). These disorders belong to a large group of diseases that are characterized by production of antibodies against the individual's own antigens. The abnormality of the thymus suggests that its control over immunological responses has broken down. Production of "forbidden clones" of cells that are unable to distinguish foreign from "self" antigens may become possible, and normal tissues may then be destroyed. Although this concept of thymus malfunction remains for the moment only a stimulating hypothesis, it does acknowledge the central role of the thymus in the development of immune responses.

From the thymus of the newborn rat
Magnification × 18,400

ACKERMAN, G. A., and HOSTETLER, J. R. Morphological studies of the embryonic rabbit thymus: The *in situ* epithelial versus the extrathymic derivation of the initial population of lymphocytes in the embryonic thymus. *Anat. Rec., 166:*27 (1970).

BURNET, M. The thymus gland. *Sci. Amer., 207:*50 (November, 1962).

CLARK, S. L., JR. Cytological evidences of secretion in the thymus. *In* The Thymus: Experimental and Clinical Studies. Ciba Foundation Symposium. G. E. W. Wolstenholme and R. Porter, editors. Boston, Little, Brown and Company (1966) p. 3.

————. The thymus in mice of strain 129/J studied with the electron microscope. *Amer. J. Anat., 112:*1 (1963).

EVERETT, N. B., and TYLER, R. W. Lymphopoiesis in the thymus and other tissues: functional implications. *Int. Rev. Cytol., 22:*205 (1967).

ITO, T., and HOSHINO, T. Fine structure of the epithelial reticular cells of the medulla of the thymus in the golden hamster. *Z. Zellforsch., 69:*311 (1966).

KALPAKTSOGLORI, P. K., YUNIS, E. J., and GOOD, R. A. The role of the thymus in development of lympho-hemopoietic tissues. The effect of thymectomy on development of blood cells, bone-marrow, spleen, and lymph nodes. *Anat. Rec., 164:*267 (1969).

LEVEY, R. H. The thymus hormone. *Sci. Amer., 211:*66 (July, 1964).

MILLER, J. F. A. P. The thymus in relation to the development of immunological capacity. *In* The Thymus: Experimental and Clinical Studies. Ciba Foundation Symposium. G. E. W. Wolstenholme and R. Porter, editors. Boston, Little, Brown and Company (1966) p. 153.

NOSSAL, G. J. V. The cellular basis of immunity. *Harvey Lect., 63:*179 (1967-68).

OSOBA, D., and MILLER, J. F. A. P. The lymphoid tissues and immune responses of neonatally thymectomized mice bearing thymus tissue in millipore diffusion chambers. *J. Exp. Med., 119:*177 (1964).

SANEL, F. T. Ultrastructure of differentiating cells during thymus histogenesis. *Z. Zellforsch. Mikros. Anat., 83:*8 (1967).

WEISS, L. Electron microscopic observations on the vascular barrier in the cortex of the thymus of the mouse. *Anat. Rec., 145:*413 (1963).

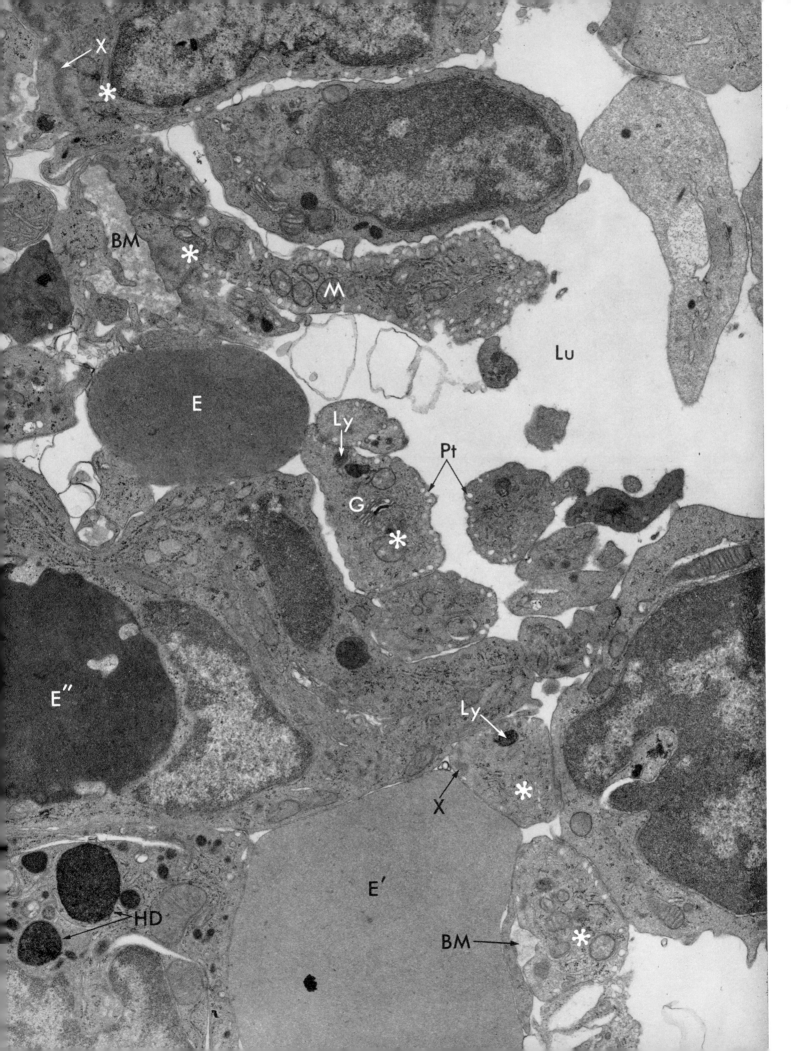

PLATE 37

Sinusoids of the Spleen

In the spleen the blood is filtered by exposure to cells belonging to the reticuloendothelial system. These are phagocytes or potential phagocytes embedded in a fine fibrous meshwork. Under the light microscope the meshwork, or reticulin, may be visualized by means of silver staining or the PAS reaction. Under the electron microscope, this extracellular material (BM) is amorphous and of medium density, and it bears a striking resemblance to the thin basement membranes underlying the various epithelia illustrated in this atlas (see, for example, Plates 10, 24–26, and 28). Occasionally fine collagen fibrils are found associated with the reticulin. As in other lymphoid tissue, such as that of the lymph nodes, this supporting network contains centers where lymphocytes and plasma cells mature. These areas in the spleen are called the white pulp. But unlike certain other lymphoid tissue, the splenic reticular tissue is infiltrated by the blood vascular system. The presence of numerous erythrocytes accounts for the appearance of the areas known as the red pulp.

While circulation in the spleen has been carefully studied, it is still uncertain just how the blood gains entrance to the reticular tissue surrounding the sinuses, the so-called cords of Billroth. It is evident that there is a complicated system of small arteries and that prominent venous sinuses are the primary collecting vessels. But it is not known whether the arteries empty into the interstices of the reticular tissue or directly into the sinuses through a closed vascular system. In any case, there are openings in the walls of the sinuses so that the movement of cells both into and out of the vascular system in these regions is a definite possibility.

In this micrograph, the wall of a venous sinus may be examined. At the right is the lumen (Lu), and portions of its lining or littoral cells (∗) are arrayed diagonally across the field. These cells differ in several respects from endothelia that line much of the vascular system. They are, first of all, much thicker (see Plates 10 and 25) and may attain a columnar shape in the region containing the nucleus (top of the plate). Their cytoplasm is rich in mitochondria (M), ribosomes are abundant, and a few Golgi regions may be identified (G). Pits (Pt) and small vesicles, like those commonly found in endothelial cells, are especially abundant at or near the free surfaces of these sinusoidal lining cells. In addition, they contain lysosomes (Ly).

The littoral cells are supported by reticulin (BM), which forms an incomplete basement membrane. An irregular band of dense material (X) lies in the basal region of the cytoplasm, facing the underlying membrane. It seems most probable that this material aids in supporting the littoral cells.

Perhaps the most striking feature of the venous sinus is the large gaps that may exist between cells. In this instance an erythrocyte (E) lies near one of these openings. In the lower part of the field, another erythrocyte (E′) seems to have passed completely out of the vascular system.

The reticuloendothelial cells have the extraordinary ability to inspect red cells and to recognize worn or damaged individuals, which they engulf and destroy. Such a red cell (E″) is present within a large reticular cell, a portion of which projects between two littoral cells. After being phagocytized, the red cell increases in density and is eroded away at the surface. As erythrocytes are broken down, bilirubin, a pigment derived from hemoglobin, is released into the blood stream. Residual matter from the red cell breakdown in the form of dense granules called hemosiderin deposits (HD) remains in the cells. The hemosiderin contains an iron-protein complex called ferritin. The iron is conserved and eventually finds its way into the hemoglobin of erythroblasts (see Plate 33).

The conditions which make worn or damaged cells subject to destruction by phagocytes are not entirely known. In the disease called hereditary spherocytosis the red cells are abnormal and are rapidly destroyed by the spleens of individuals suffering from the disease and by spleens of normal individuals receiving injections of the spherotic cells. However, in persons without spleens, the life span of the abnormal spherotic red cells is nearly normal (120 days). More recent work indicates that excessively high permeability of the red cell to sodium ion may be the primary defect in this disease.

From the spleen of the guinea pig
Magnification ×17,000

References

CROSBY, W. H. Normal functions of the spleen relative to red blood cells: A review. *Blood, 14:*399 (1959).

FELDMAN, J. D. Ultrastructure of immunologic processes. *Advances Immun., 4:*175 (1964).

JACOB, H. S., and JANDL, J. H. Increased cell membrane permeability in the pathogenesis of hereditary spherocytosis. *J. Clin. Invest., 43:*1704 (1964).

SIMON, G., and PICTET, R. Étude au microscope électronique des sinus splénique et des cordons de Billroth chez le rat. *Acta Anat., 57:*163 (1964).

THOMAS, C. E. An electron- and light microscope study of sinus structure in perfused rabbit and dog spleens. *Amer. J. Anat., 120:*527 (1967).

WEISS, L. The structure of fine splenic arterial vessels in relation to hemoconcentration and red cell destruction. *Amer. J. Anat., 111:*131 (1962).

————. Appendix. The role of the spleen in the removal of normally aged red cells. *Amer. J. Anat., 111:*175 (1962).

————. The Cells and Tissues of the Immune System. Englewood Cliffs, New Jersey, Prentice-Hall (1972).

PLATE 38
Skeletal Muscle and the Sarcoplasmic Reticulum

PLATE 38

Skeletal Muscle and the Sarcoplasmic Reticulum

THE cross striations that characterize skeletal and cardiac muscle reflect the patterned organization of contractile fibrils that fill the cytoplasm of each large, cylindrical multinucleated cell or fiber. Several of these fibrils (*) located in the marginal zone of a fiber are shown in this micrograph. They in turn are clearly constructed of filaments, and the distribution of these is related to the alternating light and dark bands. Since in separate fibrils these bands are in precise register, the total effect results in the well-known striated appearance of the cell.

Each repeating sequence of striations constitutes what is referred to as a sarcomere, and the segment thus defined is considered to be the functional unit of contraction. From among the several bands in the striation pattern, the Z line is commonly selected as marking the limits of the sarcomere. This line is unusually dense, especially in contracted fibrils, and may be correctly regarded as a kind of septum that is continuous transversely across the fibril. Other bands in the sarcomere are labeled. The isotropic band, I, is bisected by the Z. The anisotropic, A, is the more dense and is bisected by a narrow light band, the H. And frequently a line, called the M, appears along the middle of the H band.

Careful studies of the fine structure of myofibrils have revealed the presence of at least two kinds of filaments in the fibrils. Of these, the thicker and more prominent run along the length of the A band. In part they represent the protein myosin, which among other properties possesses ATP-ase activity. The second type of filamentous unit is thinner and less obvious than the first. It intermingles in a highly ordered pattern with the myosin filaments and in each sarcomere extends from the Z line to the edge of the H band. In contraction, these actin filaments slide between the myosin units with the consequent disappearance of the I and H bands. The myosin filaments have small lateral projections, which introduce a finer, transverse periodicity into the fibril and which are thought to function in the sliding motion of contraction and in binding the two filament systems together.

In addition to the fibrils, this micrograph shows a prominent vesicular component in the interfibrillar cytoplasm or, as it is called in muscle, the sarcoplasm. This is the endoplasmic reticulum of the muscle cell, referred to appropriately in muscle as the sarcoplasmic reticulum or SR. As in other cells it is made up of interconnected, anastomosing tubules and vesicles, but in this instance they are without attached ribosomes. It can thus be thought of as a smooth form of the reticulum (see Plates 2, 7, 18, and 19). There are other peculiar features about it, one being a structural pattern that repeats with respect to each sarcomere of the myofibril. Thus one can notice in the micrograph that the lacelike reticular structures (SR) are interrupted in their continuity at each Z line and seem to terminate in dilated sacs (SR'). The latter are closely apposed to an interposed vesicular element (TS), which is positioned exactly opposite the Z line. This third element of the triad (text figure 38a) is now known to be part of a transverse membrane system, called the T-system (TS), which is continuous with the sarcolemma, the limiting membrane of the muscle fiber.

The recognition of this T-system elucidated, if not solved, a basic problem of muscle contraction—that is, the surprising fact that myofibrils within the center of a fiber, perhaps 50 micrometers from the surface, contract simultaneously with fibrils at the surface and near to the electrical excitation which moves over the sarcolemma just in advance of contraction. Since diffusion rates within cytoplasm would be too slow to account for such a rapid propagation of excitation throughout the fiber, it was argued that a structural element must be involved. Therefore, against this background of physiological information and theory, the discovery of the T-system and more especially its continuity with the sarcolemma initiated a new series of investigations focused on the mechanisms of excitation-contraction coupling. Comparative morphological studies have revealed that in some muscles, particularly the more rapidly contracting ones, the T-system with associated SR vesicles is more elaborately and extensively developed and appears opposite the two halves of the I band, at the level at which A and I bands meet. Biochemical studies on isolated fragments of the SR (muscle microsomes) have shown them to possess a remarkable capacity to take up and store calcium ion

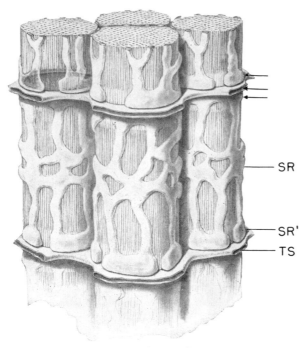

SR

SR'

TS

Text Figure 38a

This three-dimensional drawing is designed to clarify the structural relationship between the myofibrils and two smooth membrane systems, the sarcoplasmic reticulum and the T-system, found in skeletal muscle fibers. The myofibrils are cylindri-cal shafts of closely packed filaments. At the top of the drawing, four fibrils are seen cut in cross section. As described in Plate 38, the fibrils are made up of repeating units, the sarcomeres. These are drum-shaped segments, all of equal size, piled end to end, and each one is in register with those adjacent to it. Curtains of anastomosing tubules, all parts of the sarcoplasmic reticulum (SR), occupy the sarcoplasm between the myofibrils and envelop each sarcomere. The reticulum is usually expanded in the region of the I band (SR'). At the levels of the Z lines, marking the limits of the sarcomeres, the expansions lie close to the membranes of the T-system (TS), forming the so-called triad (3 arrows). The T-system represents deep, narrow infoldings of the sarcolemma or plasma membrane of the muscle fiber. Its inner space is therefore continuous with the extracellular space. There is no evidence that it ever becomes confluent with cavities of the sarcoplasmic reticulum or any other intracellular space. The T-system may be pictured as a grid perforated by pores, through which pass the myofibrils. This structure is indicated at the upper left in the drawing, where the fibril is drawn as if it were transparent. As a whole, then, the T-system resembles a series of perforated lamellae or grids, partitioning the elongated muscle cell at regular sarcomeric intervals. It is obviously in close contact with the myofibrils, as well as with the vesicles of the sarcoplasmic reticulum, providing relationships that are essential for the rapid contraction and relaxation of the muscle cell.

in the presence of ATP. More recently *in situ* tests have pinpointed the large terminal vesicles of the SR as the storage site. Presumably the excitation (depolarization) that moves over the sarcolemma preceding muscle contraction is transported internally into the fiber along the membranes of the T-system. At the triad levels this disturbance is transmitted to the adjacent SR vesicles and triggers the release of calcium ion. After initiating contraction of the adjacent myofibrils, the calcium is quickly recaptured by the SR membranes and stored for the next excitation. The relationships of these two membranous sys-tems, the SR and the T-system, to the myofibrils is illustrated further in the three-dimensional diagram in text figure 38a.

Tadpole segmental muscles are unusual in possessing very few mitochondria (sarcosomes), of which none appears in this particular field. This peculiarity reflects the fact that the muscle of origin is used by the animal only for short bursts of activity between relatively long periods of rest.

From the tail muscle of a leopard frog tadpole
Magnification × 29,000

ADELSTEIN, R. S., CONTI, M. A., JOHNSON, G. S., PASTAN, I., and POLLARD, T. D. Isolation and characterization of myosin from cloned mouse fibroblasts. *Proc. Nat. Acad. Sci. U.S.A., 69:*3693 (1972).

COSTANTIN, LeR. L., FRANZINI-ARMSTRONG, C., and PODOLSKY, J. Localization of calcium-accumulating structures in striated muscle fibers. *Science, 147:*158 (1965).

EBASHI, S., and LIPMANN, F. Adenosine triphosphate-linked concentration of calcium ions in a particulate fraction of rabbit muscle. *J. Cell Biol., 14:*389 (1962).

FAWCETT, D. W., and REVEL, J. P. The sarcoplasmic reticulum of a fast-acting fish muscle. *J. Biophysic. and Biochem. Cytol., 10:* Suppl., 89 (1961).

FISCHMAN, D. A. The synthesis and assembly of myofibrils in embryonic muscle. *Curr. Top. Devel. Biol., 5:*235 (1970).

FRANZINI-ARMSTRONG, C. Studies of the triad. III. Structure of the junction in fast twitch fibers. *Tissue and Cell, 4:*469 (1972).

————. Studies of the triad. IV. Structure of the junction in frog slow fibers. *J. Cell Biol., 56:*120 (1973).

FREYGANG, W. H. Tubular ionic movements. Symposium on excitation-contraction coupling in striated muscle. *Fed. Proc., 24:*1135 (1965).

HASSELBACH, W. Relaxation and the sarcotubular calcium pump. *Fed. Proc., 23:*909 (1964).

HEUSON-STIENNON, J-A., WANSON, J-C., and DROCHMANS, P. Isolation and characterization of the sarcoplasmic reticulum of skeletal muscle. *J. Cell Biol., 55:*471 (1972).

HUXLEY, A. F., and TAYLOR, R. E. Local activation of striated muscle fibers. *J. Physiol., 144:*426 (1958).

HUXLEY, H. E. The double array of filaments in cross-striated muscle. *J. Biophysic. and Biochem. Cytol., 3:* 631 (1957).

ISHIKAWA, H., BISCHOFF, R., and HOLTZER, H. Formation of arrowhead complexes with heavy meromyosin in a variety of cell types. *J. Cell Biol., 43:*312 (1969).

KATZ, A. M. Contractile proteins of the heart. *Physiol. Revs., 50:*63 (1970).

PEACHEY, L. D. The sarcoplasmic reticulum and transverse tubules of the frog's sartorius. *J. Cell Biol., 25* (No. 3, part 2): 209 (1965).

PORTER, K. R., and PALADE, G. E. Studies on the endoplasmic reticulum. III. Its form and distribution in striated muscle cells. *J. Biophysic. and Biochem. Cytol., 3:*269 (1957).

SMITH, D. S. Reticular organizations within the striated muscle cell. An historical survey of light microscope studies. *J. Biophysic. and Biochem. Cytol., 10:* Suppl., 61 (1961).

WANSON, J-C., and DROCHMANS, P. Role of the sarcoplasmic reticulum in glycogen metabolism. *J. Cell Biol., 54:*206 (1972).

ZADUNAISKY, J. A. The localization of sodium in the transverse tubules of skeletal muscle. *J. Cell Biol., 31:*C-11 (following p. 214) (1966).

159

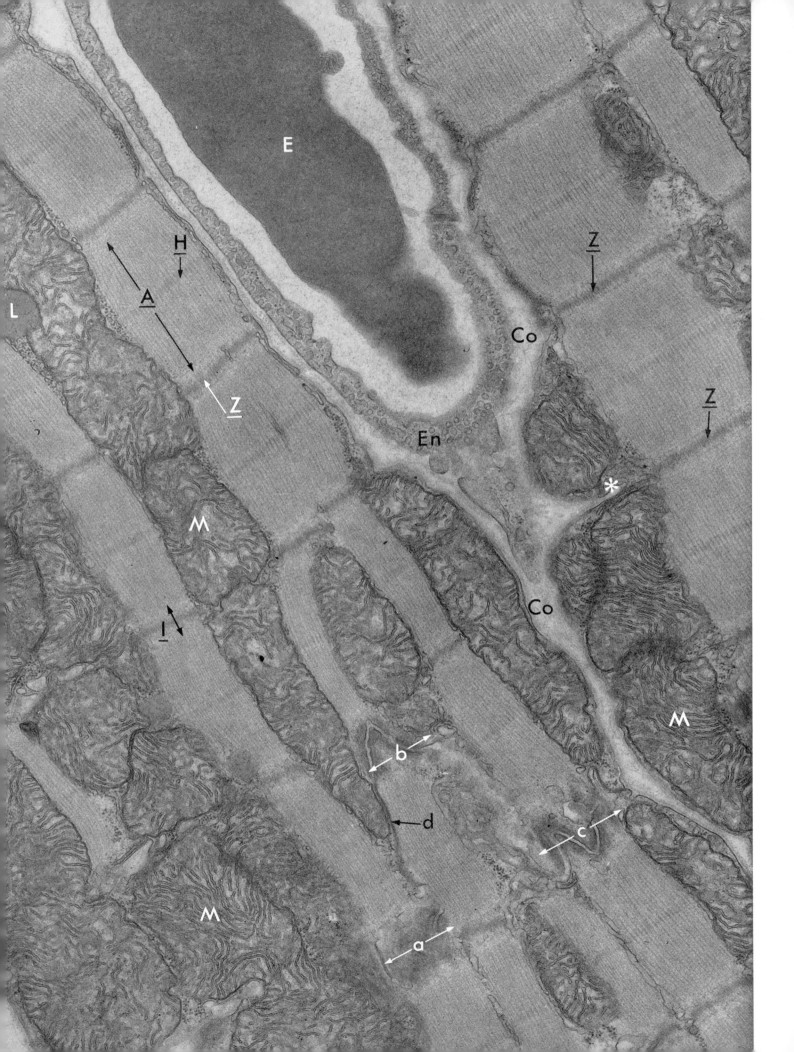

PLATE 39

Cardiac Muscle

CARDIAC muscle, like skeletal muscle, is striated, and the organization of the contractile fibrils and other components differs from that of skeletal muscle in detail only. In this micrograph portions of the cytoplasm of several cardiac muscle cells (fibers) may be examined. The cell at upper right is separated from the lateral surfaces of the column of cells at left by a thin layer of fine connective tissue fibrils (Co), within which is lodged a capillary of this well-vascularized tissue. The vesicular cytoplasm of the endothelial cells (En) and a profile of an erythrocyte (E) are evident.

As in skeletal muscle, myofibrils run parallel to the long axis of the cell, but unlike the fibrils in skeletal muscle, these may branch. The sarcomeres show the same pattern of banding as that of skeletal muscle. In this micrograph the Z lines (Z) delimiting the sarcomere are prominent and readily identified. However, the muscle is contracted so that the I band regions (I) have nearly disappeared, and the A band (A) occupies most of the sarcomere length. In the contracted state, a dark band normally appears at the position of the H band (H).

Cardiac muscle is noted for the size and the number of its mitochondria or sarcosomes (M), presumably present to satisfy the cell's unusual requirements for ATP. These are aligned in columns between the myofibrils. Often lipid droplets (L) lie alongside the sarcosomes.

For many years cardiac muscle was thought to be a syncytium. Now, however, the significance of the intercalated disk, long a puzzle to light microscopists, is understood. The disk is really an elaborate region of attachment that links the elongated cells end to end. Thus parallel columns of fibers are formed. In this micrograph two cells are joined by an intercalated disk (a, b, c). Though the disk always occurs at the level of the Z lines, it does not always occur at Z lines that are in register: i.e., the level of the disk may shift by the length of one sarcomere. Thus in this micrograph the central portion of a disk (b)

is at the level of a different Z line from those portions on either side (a, c). It follows that the boundaries of adjacent cells are not all in one plane but essentially interdigitate with one another. Along this specialized junction the plasma membranes of the attached cells can be followed, and along their cytoplasmic surfaces a dense material is accumulated. From the standpoints of morphology and function these regions within the intercalated disk may be regarded as elaborate desmosomes (see Plate 9). On the other hand, the plasma membranes (d) lateral to the sarcomere come close to one another to form a junction with a gap of about 20 A between adjacent cell membranes. When fine colloidal particles are introduced experimentally into the tissue, they can penetrate this gap, thus providing a negative stain that reveals hexagonally packed structures (each 70–75 A wide) spanning the gap. Such junctions are believed to be the sites where an electrical impulse can be transmitted from one muscle fiber membrane to the next. Thus the excitatory impulse sweeping as a wave over the plasma membrane of one fiber does not require chemical mediators for transmission to neighboring fibers (cf. Plates 42 and 43). In the propagation of the excitatory impulse that precedes and in effect causes contraction, cardiac muscle does then behave as if it were a syncytium.

The sarcolemma of the lateral surfaces of the cardiac muscle fiber is invaginated at the level of the Z line (∗). By this device the plasma membrane is brought into close contact not only with smooth-walled elements of the sarcoplasmic (endoplasmic) reticulum but also, as in the case of skeletal muscle, with the fibrils more deeply situated in the fibers. The infoldings of the sarcolemma therefore constitute the T-system of cardiac muscle (cf. Plate 38). Cardiac and skeletal muscles are therefore provided with specialized intracellular membrane systems, which are undoubtedly involved in the contractile process.

From the heart of the bat
Magnification × 30,500

DA SILVA, P. P., and GILULA, N. B. Gap junctions in normal and transformed fibroblasts in culture. *Exp. Cell Res., 71:*393 (1972).

ESSNER, E., NOVIKOFF, A. B., and QUINTANA, N. Nucleoside phosphatase activities in rat cardiac muscle. *J. Cell Biol., 25* (No. 2, part 1):201 (1965).

FAWCETT, D. W., and SELBY, C. C. Observations on the fine structure of the turtle atrium. *J. Biophysic. and Biochem. Cytol., 4:*63 (1958).

FORSSMAN, W. G., and GIRARDIER, L. A study of the T system in rat heart. *J. Cell Biol., 44:*1 (1970).

KELLY, A. M. Sarcoplasmic reticulum and T tubules in differentiating rat skeletal muscle. *J. Cell Biol., 49:* 335 (1971).

MATTER, A. A morphometric study on the nexus of rat cardiac muscle. *J. Cell Biol., 56:*690 (1973).

McNUTT, N. S. Ultrastructure of intercellular junctions in adult and developing cardiac muscle. *Amer. J. Cardiol., 25:*169 (1970).

————, and WEINSTEIN, R. S. The ultrastructure of the nexus: a correlated thin-section and freeze-cleave study. *J. Cell Biol., 47:*173 (1971).

NELSON, D. A., and BENSON, E. S. On the structural continuities of the transverse tubular system of rabbit and human myocardial cells. *J. Cell Biol., 16:*297 (1963).

PAPPAS, G. D., ASADA, Y., and BENNETT, M. V. L. Morphological correlates of increased coupling resistance at an electrotonic synapse. *J. Cell Biol., 49:*173 (1971).

REVEL, J. P., and KARNOVSKY, M. J. Hexagonal array of subunits in intercellular junctions. *J. Cell Biol., 33:* C-7 (following p. 450) (1967).

SIMPSON, F. O., and OERTELIS, S. J. The fine structure of of sheep myocardial cells; sarcolemmal invaginations and the transverse tubular system. *J. Cell Biol., 12:*91 (1962).

SJÖSTRAND, F. S., and ANDERSSON-CEDERGREN, E. Intercalated discs of heart muscle. *In* The Structure and Function of Muscle. G. H. Bourne, editor. New York, Academic Press, vol. I (1960) p. 421.

PLATE 40

Smooth Muscle

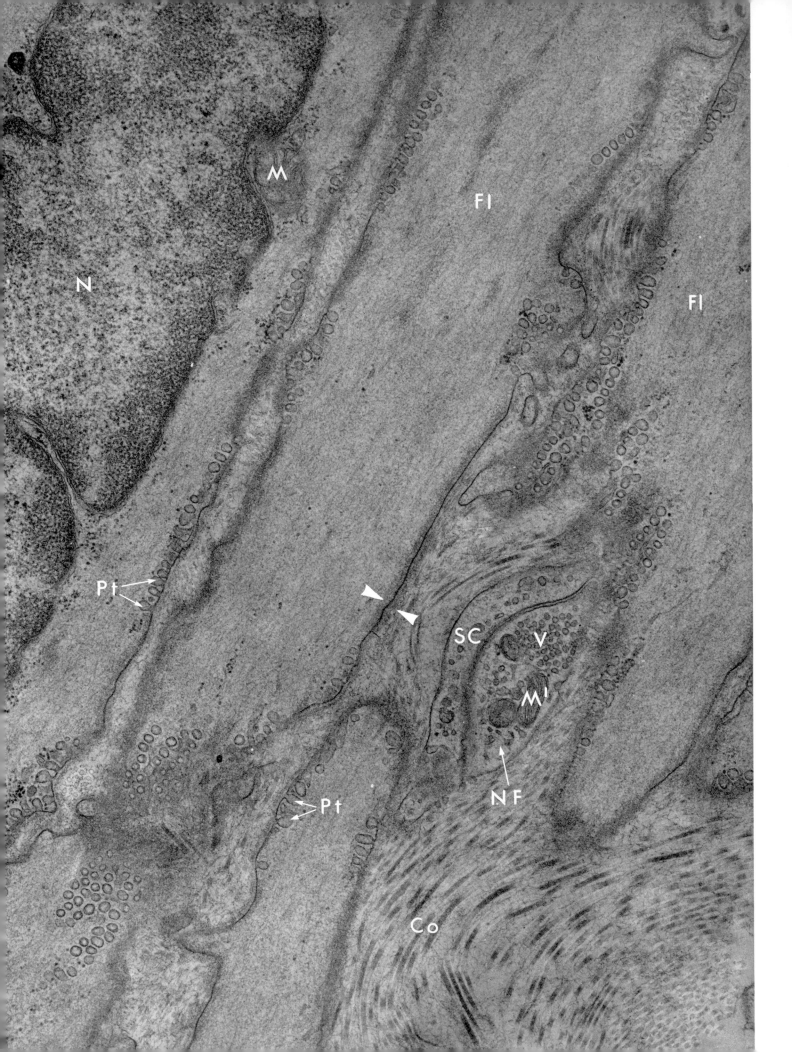

PLATE 40

Smooth Muscle

Smooth muscles, which contract involuntarily, are found in numerous places in the vertebrate body, such as in the walls of the intestinal tract, the blood vessels, and the uterus. They are characterized by several physiological features, of which their slow contraction time is outstanding. In some instances seconds, or even minutes, are required to complete a single contraction.

The tissue itself is made up of spindle-shaped cells containing a system of contractile proteins. The cells are intimately associated with the connective tissue fibers that bind them together. In these respects smooth muscle is similar to striated muscle, but unlike the latter the contractile elements are not organized into any obvious pattern for which microscopists have been able to provide a satisfactory functional interpretation (cf. Plates 38 and 39). Very thin filaments (Fl) can be seen to run parallel to the long axis of the cell (fiber) and to constitute the bulk of the cytoplasm. Thick filaments, common in striated fibers and identifiable with the protein myosin, are not obvious and have been observed only rarely in smooth muscle cells. Mitochondria (M) usually cluster near the irregular border of the nucleus (N). The endoplasmic reticulum is represented only by a few smooth-surfaced vesicles lying in the perinuclear region and in sparsely distributed channels among the fibrous elements.

Aside from the abundance of contractile material, specialized structural features associated with the plasma membrane distinguish smooth muscle cells. Representative of these are certain flasklike invaginations (pits) of the membrane (Pt) lying in the peripheral cytoplasm. The form of these suggests a structural device by which substances may be incorporated into the cells by a process akin to pinocytosis. Until now, however, evidence of uptake by these vesicles is lacking, and there is certainly no reason to believe that they transport selected molecules across the muscle fibers as they may across cells of the vascular endothelium, where they occur in similar form and number (cf. Plate 10).

Also associated with the plasma membrane are fine feltworks of moderately dense material lying on opposite sides of the line representing the membrane (arrows). Of these, the extracellular material is closely interwoven with the matrix and fibrils of the surrounding connective tissue, and it seems probable that the feltwork is homologous with the basement membranes (basal laminae) of other tissues (see Plates 10, 25, and 28). As such it serves to bind together the muscle cells and connective tissue fibers such as those seen at lower right (Co).

Studies of smooth muscle have revealed that, as in the case of striated muscles, actin and myosin are both present as the contractile proteins. Surprisingly they are similar in most respects to their counterparts in striated muscle, even though their disposition within the cell is strikingly different. The thin filaments, already mentioned, correspond to actin and may be isolated readily in filamentous form. The thick filaments, which are seemingly rare, probably represent myosin, but thus far this protein has been isolated as filaments only under rather special experimental conditions. It appears that most of the myosin in smooth muscle cells exists in a disaggregated form. Since the diameter of the thin filaments does not change measurably during contraction, models have been proposed in which myosin dimers act as lateral bridges that join arrays of actin filaments sliding past each other during contraction.

Comparison of relaxed and contracted smooth muscle indicates some of the morphological events accompanying shortening. Thin filaments become straighter and more clearly organized into bundles. At the same time areas of the plasma membrane lined with a dense layer of material (left arrow) fold inward as if the thin filaments were attached to and exerting a force on them. As though to emphasize this, intervening areas of the plasma membrane, where pits (Pt) are abundant, bulge outward. The strandlike cytoplasmic densities, called dense bodies (above label Fl), become longer. Like the cortical dense material, they also represent regions where actin filaments insert or are anchored. Thus these intracellular densities and those associated with the plasma membrane correspond to the Z lines of the striated muscle fibril. As contraction occurs, the nuclear envelope also becomes folded, and the mitochondria, ribosomes, etc., are aggregated at the nuclear poles.

Recently there have been observed special regions of attachment in which plasma membranes of adjacent smooth muscle fibers are closely associated, and no connective tissue fibers intervene between them. In some instances only 20 A separates the outer leaflets of the unit membranes, and thus a gap junction is formed uniting adjacent cells. In others the contact is less intimate and a space of ~100 A persists. It has been proposed that such regions of contact, particularly the gap junctions, may function in the transmission of excitation (the stimulation to contract) from one fiber to another.

A cross section of a small autonomic nerve process (NF) lies near the muscle fiber on the right-hand side of the picture. Mitochondria (M')

and vesicles (V), both of which are found near the regions of the synapse between nerve and muscle (or nerve), indicate that this is a terminal portion of a nerve fiber. However, an actual synapse, represented by the close approximation of two plasma membranes (see Plate 43), is not included in the section. The cytoplasm (SC) covering one side of the nerve fiber belongs to a Schwann cell. The structure of nervous tissue is discussed in details in Plates 42 through 45.

From the esophagus of the bat
Magnification × 38,500

References

BURNSTOCOK, G. Structure of smooth muscle and its innervation. *In* Smooth Muscle. E Bülbring, A. F. Brading, A. W. Jones, and T. Tomita, editors. Baltimore, Williams and Wilkins Company (1970) p. 1.

CAESAR, R., EDWARDS, G. A., and RUSKA, H. Architecture and nerve supply of mammalian smooth muscle tissue. *J. Biophysic. and Biochem. Cytol., 3:*867 (1957).

COOKE, P. H., and FAY, F. S. Correlation between fiber length, ultrastructure, and the length-tension relationship of mammalian smooth muscle. *J. Cell Biol., 52:*105 (1972).

DEWEY, M. M., and BARR, L. Intercellular connection between smooth muscle cells: the nexus. *Science, 137:* 670 (1962).

ELLIOTT, G. F. X-ray diffraction studies on striated and smooth muscles. *Proc. Roy. Soc., Series B, 160:*467 (1964).

————. Variations of the contractile apparatus in smooth and striated muscles. X-ray diffraction studies at rest and in contraction. *J. Gen. Physiol., 50* (Suppl.):171 (1967).

FAY, F. S., and COOKE, P. H. Reversible disaggregation of myofilaments in vertebrate smooth muscle. *J. Cell Biol., 56:*399 (1973).

KELLY, R. E., and RICE, R. V. Localization of myosin filaments in smooth muscle. *J. Cell Biol., 37:*105 (1968).

————, and RICE, R. V. Ultrastructural studies on the contractile mechanism of smooth muscle. *J. Cell Biol., 42:*683 (1969).

LANE, B. P. Alterations in the cytologic detail of intestinal smooth muscle cells in various stages of contraction. *J. Cell Biol., 27:*199 (1965).

————, and RHODIN, J. A. G. Cellular interrelationships and electrical activity in two types of smooth muscle. *J. Ultrastruct. Res., 10:*470 (1964).

LOWY, J., and SMALL, J. V. The organization of myosin and actin in vertebrate smooth muscle. *Nature, 227:*46 (1970).

OOSAKI, T., and ISHII, S. Junctional structure of smooth muscle cells. The ultrastructure of the regions of junction between smooth muscle cells in rat small intestine. *J. Ultrastruct. Res., 10:*567 (1964).

PANNER, B. J., and HONIG, C. R. Filament ultrastructure and organization in vertebrate smooth muscle. Contraction hypothesis based on localization of actin and myosin. *J. Cell Biol., 35:*303 (1967).

————, and HONIG, C. R. Locus and state of aggregation of myosin in tissue sections of vertebrate smooth muscle. *J. Cell Biol·, 44:*52 (1970).

RICE, R. V., McMANUS, G. M., DEVINE, C. E., and SOMLYO, A. P. Regular organization of thick filaments in mammalian smooth muscle. *Nature New Biol., 231:*242 (1971).

SHOENBERG, C. F. An electron microscope study of the influence of divalent ions on myosin filament formation in chicken gizzard extracts and homogenates. *Tissue and Cell, 1:*83 (1969).

————, RÜEGG, J. C., NEEDHAM, D. M., SCHIRMER, R. H., and NEMETCHEK-GANSLER, H. A biochemical and electron microscope study of the contractile proteins in vertebrate smooth muscle. *Biochem. Z., 345:* 255 (1966).

UEHARA, Y., CAMPBELL, G. R., and BURNSTOCK, G. Cytoplasmic filaments in developing and adult vertebrate smooth muscle. *J. Cell Biol., 50:*484 (1971).

PLATE 41

The Arteriole

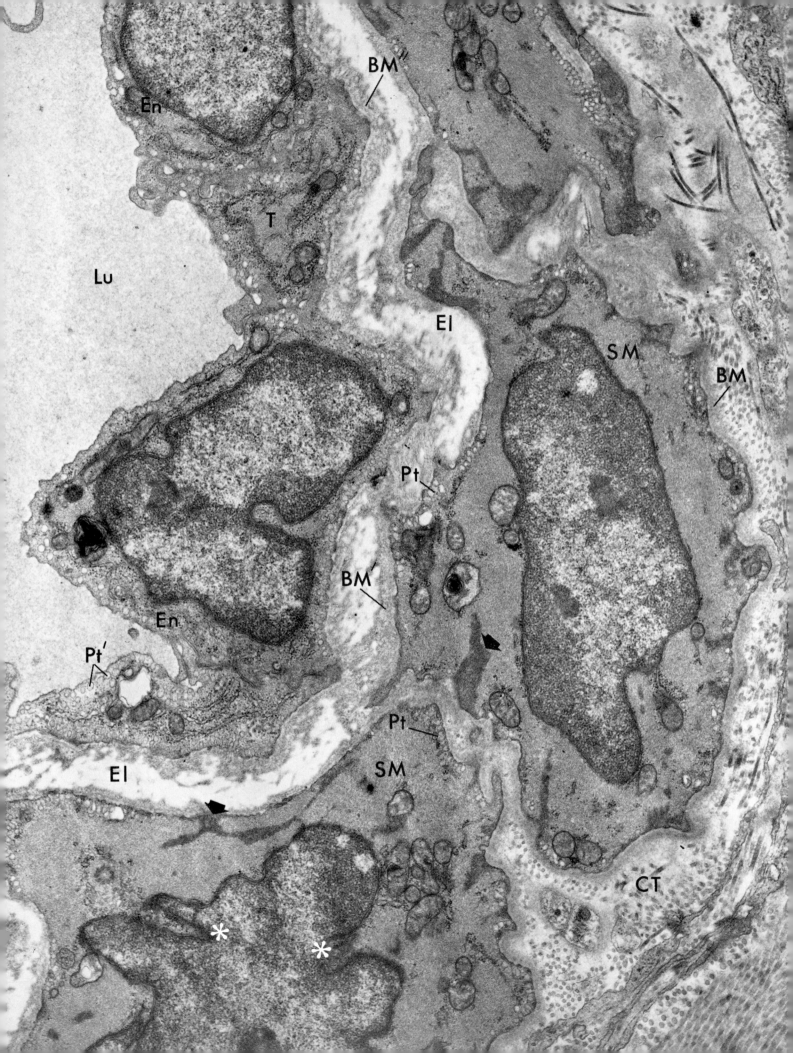

PLATE 41

The Arteriole

THE terminal arterioles, among the smallest branches of the arterial tree, are part of the strong-walled muscular system of vessels that conduct and distribute the blood to the capillary beds. They receive blood from the larger arteries that carry it from the heart. From them in turn the blood passes through the precapillary sphincters, then into small vessels called metarterioles, and finally into the capillaries. In contrast with capillaries, the importance of the arterioles and the sphincters lies not in their ability to bring about exchanges of metabolites between the capillary lumen and the surrounding tissues, but rather in their effectiveness in regulating blood pressure. The caliber of the arterioles changes in response to nervous stimulation and to changes in the composition of the blood. They thus serve as a kind of dam and floodgate, helping to regulate the flow of blood into the capillaries. Since capillary beds, such as those in the abdominal viscera, may accommodate the entire normal volume of blood, general dilation of arterioles that lead to these capillaries would result (as it does in certain cases of shock following sudden trauma) in a precipitous drop in blood pressure. Under conditions that usually prevail, the narrowing of the arterioles makes an important contribution to the maintenance of normal blood pressure and the allocation of blood to the various capillary beds.

The wall of a terminal arteriole, which may be examined in this micrograph, is made up of an endothelium (En) lining the lumen (Lu) of the vessel, a layer of elastic connective tissue (El), a circumferentially arranged layer of smooth muscle (SM), and a peripheral layer of collagenous connective tissue (CT).

It is, of course, the contraction and relaxation of the smooth muscle that controls directly the size of the arteriole lumen. The terminal arteriole has been defined as a vessel having a diameter of less than 50 μm and having a complete muscular layer one cell thick. The muscle cell layer forms the sphincter at the bases of smaller (less than 15 μm in diameter) vessels branching from the arterioles and then gradually disappears as the vessels (metarterioles) lead into arterial capillaries.

The muscle fibers seen here in cross section are in a contracted state, as evidenced by the irregu-

lar folds of the nuclear envelope (*) and the outward bulging of the plasma membrane in regions containing many pits (Pt). The latter regions alternate with infoldings where dense bodies (arrows) are contiguous with the inner surface of the plasma membrane (cf. Plate 40). The basal lamina (BM) surrounding the muscle fibers is more compact on the sides of the cells facing the collagenous connective tissue layer (CT) or adventitia. Careful examination of the small arterioles has demonstrated that there are abundant nerve endings associated with muscle cells. Both vasoconstrictor and vasodilator nerve fibers are found, but the former are believed to be of prime importance. Furthermore, gap junctions between smooth muscle cells also occur regularly. These junctions probably facilitate the spread of the excitatory impulse for contraction from muscle cell to muscle cell.

Elastic fibers (El), which are unstained in this preparation, form a thin connective tissue layer that is enmeshed with basement membrane material, both with that surrounding the smooth muscle (BM′) and that underlying the endothelium (BM″). Elastic fibers are stretched as the vessel is distended and due to their elasticity add to the force exerted by contracting muscle fibers in narrowing the vessel lumen. In the large arteries, elastic connective tissue is extensively developed. There its property of resisting deformation is important in helping to maintain tension of the arterial wall.

The endothelium (En) lining the terminal arteriole is a thick layer when compared to many capillary linings (see Plates 10 and 25). The cells making it up are probably capable of expanding and contracting passively as the diameter of the lumen varies. Bundles of tonofilaments (T) may strengthen this layer. Cells of the endothelium are joined closely together, and fenestrae are lacking. As in other endothelial cells, pits (Pt′) and vesicles derived from them are prominent structural features.

In the terminal arterioles and precapillary sphincters there exist special areas of contact between endothelial cells and muscle fibers, such as those in text figure 41a (X). Tongues of endothelial cell cytoplasm extend through the underlying basement membrane to form myoendo-

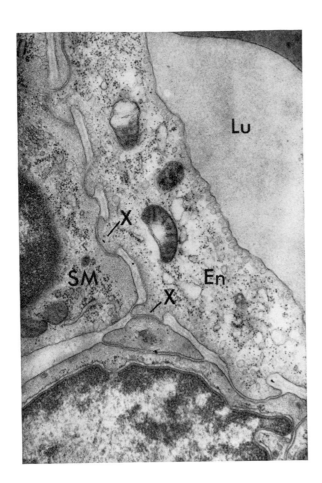

Text Figure 41a

Endothelial cells of terminal arterioles and pre-capillary sphincters, such as that shown here (En), frequently have footlike processes (X), and each makes a special contact with smooth muscle fibers (SM) of the vessel wall. The processes extend through the basement membrane material separating the two types of tissue. At the region of contact with the smooth muscle cell an intercellular gap of only 45 A remains. Such myoendothelial junctions may facilitate transmission of humoral agents from the blood in the vessel lumen (Lu) to muscle fibers, thus helping to regulate the dilation and constriction of the smallest arterial vessels. Micrograph courtesy of Dr. J. A. G. Rhodin. It is reprinted from Figure 24, *J. Ultrastruct. Res., 18*:181 (1967).

From the thigh muscles of the rabbit (*Oryctolagus cuniculus*)
Magnification × 25,500

thelial junctions with smooth muscle fibers. The two cells are separated by a 45 A gap but not by intervening basement membrane material.

Observations on small arterial vessels deprived of nervous connections have indicated that in some cases, such as in skeletal muscle, a strong muscular tone remains. These results have been interpreted as an indication that substances present locally can control the vasomotor responses. The presence of myoendothelial junctions may facilitate transmission of humoral substances from the blood to muscle fibers. Possibly for sub-stances diffusing across the endothelium the junctions serve as pathways direct to the muscle. Perhaps more intriguing is the speculation that certain of the endothelial cells may act as receptors that respond to local concentrations of agents in the blood by depolarization of the plasma membrane. This excitation would then be transmitted to the muscle membrane at the junctional contacts.

From the epididymis of the hamster (*Cricetus cricetus*)
Magnification × 21,500

FAWCETT, D. W. The fine structure of capillaries, arterioles and small arteries. *In* The Microcirculation. S. R. M. Reynolds and B. W. Zweifach, editors. Urbana, Ill., University of Illinois Press (1959) p. 1.

GREENLEE, T. K., JR., ROSS, R., and HARTMAN, J. L. The fine structure of elastic fibers. *J. Cell Biol., 30:*59 (1966).

HÜTTNER, I., BOUTET, M., and MORE, R. H. Gap junctions in arterial endothelium. *J. Cell Biol., 57:*247 (1973).

IWAYAMA, T. Nexuses between areas of the surface membrane of the same arterial smooth muscle cell. *J. Cell Biol., 49:*521 (1971).

KARRER, H. E. Electron microscope study of developing chick embryo aorta. *J. Ultrastruct. Res., 4:*420 (1960).

RHODIN, J. A. G. The ultrastructure of mammalian arterioles and precapillary sphincters. *J. Ultrastruct. Res., 18:*181 (1967).

RICHARDSON, J. B., and BEAULNES, A. The cellular site of action of angiotensin. *J. Cell Biol., 51:*419 (1971).

ROSS, R. The elastic fiber. A review. *J. Histochem. Cytochem., 21:*199 (1973).

SCHWARTZ, S. M., and BENDITT, E. P. Studies on aortic intima. *Amer. J. Pathol., 66:*241 (1972).

STEHBENS, W. E., and SILVER, M. D. Unusual development of basement membrane about small blood vessels. *J. Cell Biol., 26:*669 (1965).

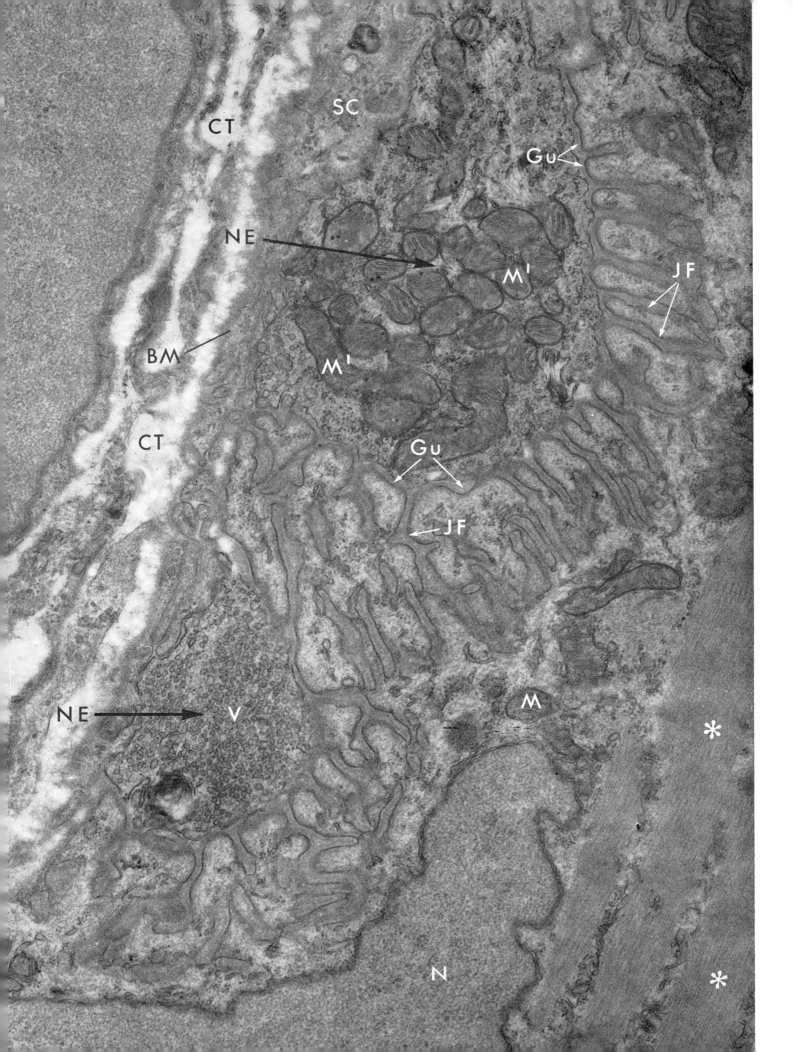

PLATE 42

The Myoneural Junction

THE stimulus to contract is brought to the fibers of skeletal muscle by axon processes of motor neurons, and the region of contact between the two cell types is especially differentiated to achieve the transmission. In the accompanying micrograph several myofibrils (*) identify the striated muscle fiber (cell). Where the axon terminal reaches the muscle, it inhabits a depression called a trough or gutter (Gu), which indents the surface of the fiber. Deep infoldings of the sarcolemma, called junctional folds (JF), extend from the bottoms of the troughs into the underlying cytoplasm, or sole plasm, of the fiber. Mitochondria (M) and the nucleus (N) of the muscle cell occur typically in this region of the fiber. The subneural apparatus, i.e., the gutters and junctional folds, are filled with a moderately dense, amorphous material that is similar to and continuous with that covering other regions of the muscle surface; it is essentially a basement membrane. The nerve cell endings (NE), on the other hand, lie naked in the trough; no cellular sheath is interposed between them and the muscle fiber. Rather, the two cells are separated from each other only by the layer of amorphous material described above. The Schwann cell sheath that encloses the axon as it approaches the myoneural junction pulls away from the axon terminal and remains only as a lidlike cover over the junction area. A portion of a Schwann cell (SC) can be seen in this illustration. It is surrounded by a basement membrane (BM) and connective tissue fibers (CT) of the endomysium. These structural relationships are diagrammed in text figure 42a.

At their tips, axons are filled with many small synaptic vesicles (V) and abundant mitochondria (M′). Because these vesicles are constantly present near the region of contact between nerve and muscle, as well as near synaptic junctions in the nervous system (see Plate 43), they have aroused considerable interest. Out of this has grown the theory that such vesicles, which measure 30 to 40 nm in diameter, contained the humoral transmitter, acetylcholine. This localization has been established by chemical analysis of isolated synaptic vesicles. Cholinacetylase, an enzyme necessary for the formation of acetylcholine, is probably closely associated with the vesicles as well. Presumably the release of the transmitter substance through the presynaptic membrane and into the intercellular subneural space would bring about permeability changes and subsequently changes in the electrical potential across the postsynaptic membrane of the muscle. The resulting excitation and its propagation along the muscle fiber leads finally to the contraction of the myofibrils.

Once acetylcholine has served its function, it is rapidly destroyed by the enzyme cholinesterase. Histochemical studies of isolated synaptic membranes and use of autoradiographic techniques adapted for electron microscopy indicate that the enzyme is localized in the junctional folds and is probably an integral part of the postsynaptic membrane. The junctional folds provide a sixfold increase in the surface of the muscle fiber in the junctional region. This specialization would therefore seem to ensure high concentration of enzyme and prompt breakdown of acetylcholine, thus preventing continuous action of the neurohumor. Due to the extreme rapidity of this reaction the muscle fiber is almost immediately ready to be stimulated once more.

From rat diaphragm
Magnification × 32,000

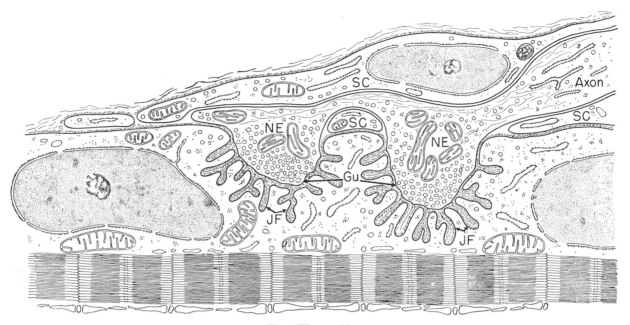

Text Figure 42a

This line drawing shows diagrammatically the relationship of the muscle fiber, nerve ending, and Schwann cell within the myoneural junction. It is based on a diagram by Couteaux, 1960.

A small part of a muscle fiber is shown in the lower half of the figure, identified by a myofibril. Nuclei lie right and left of the troughs or gutters (Gu), which indent the surface of the muscle cell. In the region of the troughs, the sarcolemma or plasma membrane is thrown into folds. These are called junctional folds (JF), and the material lining them is continuous with that covering the entire sarcolemma. This coating resembles a basement membrane and separates the muscle fiber from the nerve endings (NE) that reside in the gutters. As the axon reaches the region of the myoneural junction, its myelin sheath terminates, so that the neurolemma has no extraneous coverings except that represented by a Schwann cell (SC), which forms a lid enclosing the region of contact between the muscle and nerve cell. The axoplasm of the nerve endings is dominated by large numbers of synaptic vesicles, which crowd toward the zone of contact between the plasma membrane of the ending and the sarcolemma limiting the gutters.

References

CECCARELLI, B., HURLBUT, W. P., and MAURO, A. Depletion of vesicles from frog neuromuscular junctions by prolonged tetanic stimulation. *J. Cell Biol., 54:*30 (1972).

COUTEAUX, R. Motor end-plate structure. *In* The Structure and Function of Muscle. G. H. Bourne, editor. New York, Academic Press, vol. I (1960) p. 337.

CSILLIK, B., and KNYIHÁR, E. On the effect of motor nerve degeneration on the fine-structural localization of esterases in the mammalian motor end-plate. *J. Cell Sci., 3:*529 (1968).

DAVIS, R., and KOELLE, G. B. Electron microscopic localization of acetylcholinesterase and nonspecific cholinesterase at the neuromuscular junction by the gold-thiocholine and gold-thiolacetic acid methods. *J. Cell Biol., 34:*157 (1967).

DE ROBERTIS, E. Ultrastructure and cytochemistry of the synaptic region. *Science, 156:*907 (1967).

———, RODRIGUES DE LORES ARNAIZ, G., SALGANICOFF, L., PELLEGRINO DE IRALDI, A., and ZIEHER, L. M. Isolation of synaptic vesicles and structural organization of the acetylcholine system within brain nerve endings. *J. Neurochem., 10:*225 (1963).

MILEDI, R., and SLATER, C. R. Electrophysiology and electron-microscopy of rat neuromuscular junctions after nerve degeneration. *Proc. Roy. Soc. London, Ser. B., 169:*289 (1968).

PADYKULA, H. A., and GAUTHIER, G. F. The ultrastructure of the neuromuscular junctions of mammalian red, white, and intermediate skeletal muscle fibers. *J. Cell Biol., 46:*27 (1970).

SALPETER, M. M. Electron microscope radioautography as a quantitative tool in enzyme cytochemistry. The distribution of acetylcholinesterase at motor end plates of a vertebrate twitch muscle. *J. Cell Biol., 32:*379 (1967).

ZACKS, S. I., and SAITO, A. Uptake of exogenous horseradish peroxidase by coated vesicles in mouse neuromuscular junctions. *J. Histochem. Cytochem., 17:*161 (1969).

PLATE 43

The Motor Neuron of the Spinal Cord

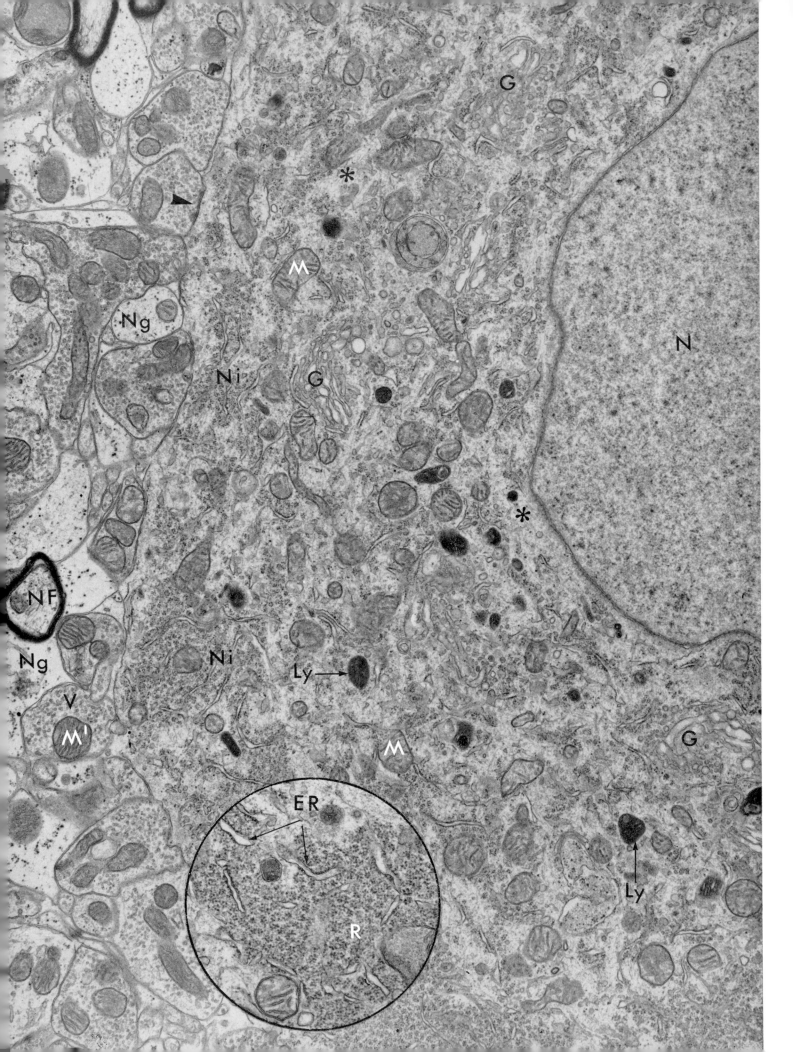

PLATE 43

The Motor Neuron of the Spinal Cord

THE motor neurons of the spinal cord, like neurons in general, are large cells. Only a small part of their total size is represented by the nucleus and perikaryon (the cytoplasm of the cell body), the greater part being in their processes, the axons and dendrites. Despite this unusual fact, the perikaryon is essential to the continuing function of the entire cell. It is the site of synthesis of the axon cytoplasm and that of the dendrites as well, and regeneration of nerves depends on the intact functioning of this central structure.

This micrograph depicts about half of a cross section through the perikaryon of a motor neuron and half its large nucleus (N). The perikaryon is rich in organelles, which in their morphology show variations on the themes displayed in other cells. The mitochondria (M) are small, fairly numerous, and more or less evenly distributed. Lysosomes (Ly) are commonly present. The Golgi region (G), which in its entirety is a complicated reticular structure in these cells, appears in section as scattered profiles of stacked cisternae and vesicles. Masses of ribosomes interlaced with profiles of ER cisternae represent the basophilic Nissl bodies (Ni), which are typical of neuronal cytoplasm. As shown in the inset, the bulk of the ribosomes (R) are not attached to ER membrane surfaces (ER), and presumably this distribution indicates that they are most active in the synthesis of proteins for retention by the neuron or its processes. Other intervening regions of the cytoplasm appear less dense than the Nissl material and can be seen in favorable instances to contain fine filaments, neurofilaments (*), which are especially common in axoplasm (see also Plates 44 and 45).

While the functional significance of the various components of the neuron cytoplasm (the perikaryon) can be surmised from generalizations based on our knowledge of cell fine structure, the specific and possibly special roles they perform in nerve cells are not as yet completely known. Only fragmentary but interesting observations are available. For example, the rough-surfaced ER, which is part of Nissl bodies, has been shown to be rich in cholinesterase. Presumably this enzyme, synthesized and sequestered in the ER, is thence transported to sites of ac-

tivity near the cell surface (cf. Plate 42). The lysosomes, also prominent in neurons, have been demonstrated to be rich in acid phosphatase. But less is known about the special roles of the masses of ribosomes also in the Nissl bodies, the neurofilaments in the cytoplasmic ground substance, and the Golgi, a vast and complex component of these cells. Evidence from autoradiographic studies suggests that in the neuron the Golgi is not involved in the packaging of a product for secretion, but its importance to the neuron has not been discovered.

We have noted before that in some instances functional information about cell organelles can be gleaned from exposing the cell to some well-defined physiological stress. When, for example, a motor neuron is deprived of its axon, it undertakes at once to regenerate a new one. As part of this phenomenon the Nissl bodies seem to lose their intense staining properties and are said to undergo chromatolysis (cf. text figure 4a). At the fine structural level the arrays of ER cisternae are seen to separate and to become fenestrated. The system of ribosomes and cisternae representing the Nissl body opens up and is essentially diluted by other cytoplasmic components with a consequent loss in staining intensity. It is apparent that the protein-synthesizing machinery of the cell (ER and ribosomes) is dramatically affected by a situation that calls for a greater than normal production of new axoplasm in nerve regeneration. One can properly infer, therefore, that the Nissl is ordinarily involved in the slow but constant replacement of neuron cytoplasm.

In the spinal cord tissue just outside the neuron body, many nerve fibers can be seen in cross section. Of these some are myelinated (NF), that is, covered by a sheath, but others are not. The derivation and structure of myelin is discussed in relation to Plate 44. The structures most closely applied to the neuron surface represent nerve terminals. These are characterized by numerous synaptic vesicles (V) and a few mitochondria (M'). In the area of functional contact between neuronal elements, that is, in the synaptic region (arrowhead), the plasma membranes of the fiber and the neuron are intact and separated from each other by a space of about 80 A. Each appears denser, however, than in nonsyn-

177

aptic areas. Because impulse conduction is uni-directional (in this case from fiber to cell body), the synapse plays a crucial role in integrating activities of the nervous system.

This synapse is one in which transmission is chemically mediated; that is, a substance liberated from the nerve ending of one cell brings about excitation in the plasma membrane of the next. In many cases acetylcholine fulfills this function just as it does in the myoneural junction (see Plate 42). In other instances norepinephrine plays a similar role, although in these instances some structural differences in the synapse are found. Transmission of nerve impulses without chemical mediation—electrical transmission—has also been detected. In these instances specialized low-resistance connections exist, coupling the pre- and postsynaptic neurons and resulting in ex-

tremely rapid transmission. In all cases in which electrical transmission has been discovered a particular structural type of intercellular junction has also been present. This resembles a "tight" junction in which the adjacent plasma membranes are fused (cf. text figure 7a).

In the central nervous system the nerve cells and their processes are supported and perhaps nourished by neuroglial cells. The small sections of membrane-limited cytoplasm (Ng) containing an occasional mitochondrion, sparsely scattered filamentous material, small vesicles of the endoplasmic reticulum, and darkly staining glycogen particles probably belong to these neuroglial elements.

From the spinal cord of the bat
Magnification × 16,000

References

BENNETT, M. V. L., ALJURE, E., NAKAJIMA, Y., and PAPPAS, G. D. Electrotonic junctions between teleost spinal neurons: electrophysiology and ultrastructure. *Science, 141:*262 (1963).

BLOOM, F. E. Ultrastructural identification of catecholamine-containing central synaptic terminals. *J. Histochem. Cytochem., 21:*333 (1973).

BODIAN, D., and MELLORS, R. C. The regenerative cycle of motoneurons with special reference to phosphatase activity. *J. Exp. Med., 81:*469 (1945).

BUNGE, M. B. Fine structure of nerve fibers and growth cones of isolated sympathetic neurons in culture. *J. Cell Biol., 56:*713 (1973).

————, BUNGE, R. P., PETERSON, E. R., and MURRAY, M. R. A light and electron microscope study of long-term organized cultures of rat dorsal root ganglia. *J. Cell Biol., 32:*439 (1967).

COTMAN, C. W., and TAYLOR, D. Isolation and structural studies on synaptic complexes from rat brain. *J. Cell Biol., 55:*696 (1972).

DEITCH, A. D., and MOSES, M. J. The Nissl substance of living and fixed spinal ganglion cells. II. An ultraviolet absorption study. *J. Biophysic. and Biochem. Cytol., 3:*449 (1957).

DE ROBERTIS, E. Ultrastructure and cytochemistry of the synaptic region. *Science, 156:*907 (1967).

————. Molecular biology of synaptic receptors. *Science, 171:*963 (1971).

ECCLES, J. The synapse. *Sci. Amer., 212:*56 (January, 1965).

FURSHPAN, E. J. "Electrical transmission" at an excitatory synapse in a vertebrate brain. *Science, 144:*878 (1964).

GRAY, E. G. The fine structural characterization of different types of synapse. *Progr. Brain Res., 34:*149 (1971).

HOLTZMAN, E. Lysosomes in the physiology and pathology of neurons. *In* Lysosomes in Biology and Pathology, Volume I. J. T. Dingle and H. B. Fell, editors.

Amsterdam, North-Holland Publishing Company (1969) p. 192.

————. Cytochemical studies of protein transport in the nervous system. *Phil. Trans. Roy. Soc. London, Ser. B., 261:*407 (1971).

————, NOVIKOFF, A. B., and VILLAVERDE, H. Lysosomes and GERL in normal and chromatolytic neurons of the rat ganglion nodosum. *J. Cell Biol., 33:*419 (1967).

PALAY, S. L. The morphology of synapses in the central nervous system. *Exp. Cell Research, Suppl. 5:*275 (1958).

————. Principles of cellular organization in the nervous system. *In* Neurosciences: A study program. G. C. Quarton, T. Melnechuk, and F. O. Schmitt, editors. New York, The Rockefeller University Press (1967) p. 24.

————, and PALADE, G. E. The fine structure of neurons. *J. Biophysic. and Biochem. Cytol., 1:*69 (1955).

PANNESE, E. Investigations on the ultrastructural changes of the spinal ganglion neurons in the course of axon regeneration and cell hypertrophy. I. Changes during axon regeneration. *Z. Zellforsch., 60:*711 (1963).

PETERS, A., PALAY, S. L., and WEBSTER, H. DEF. The Fine Structure of the Nervous System. New York, Hoeber Medical Division, Harper and Row (1970).

PRICE, D. L., and PORTER, K. R. The response of ventral horn neurons to axonal transection. *J. Cell Biol., 53:*24 (1972).

ROBERTSON, J. D., BODENHEIMER, T. S., and STAGE, D. E. The ultrastructure of Mauthner cell synapses and nodes in goldfish brains. *J. Cell Biol., 19:*159 (1963).

TEICHBERG, S., and HOLTZMAN, E. Axonal agranular reticulum and synaptic vesicles in cultured embryonic chick sympathetic neurons. *J. Cell Biol., 57:*88 (1973).

ZAMPIGHI, G., and ROBERTSON, J. D. Fine structure of the synaptic discs separated from the goldfish medulla oblongata. *J. Cell Biol., 56:*92 (1973).

178

PLATE 44

Peripheral Nerve Fibers

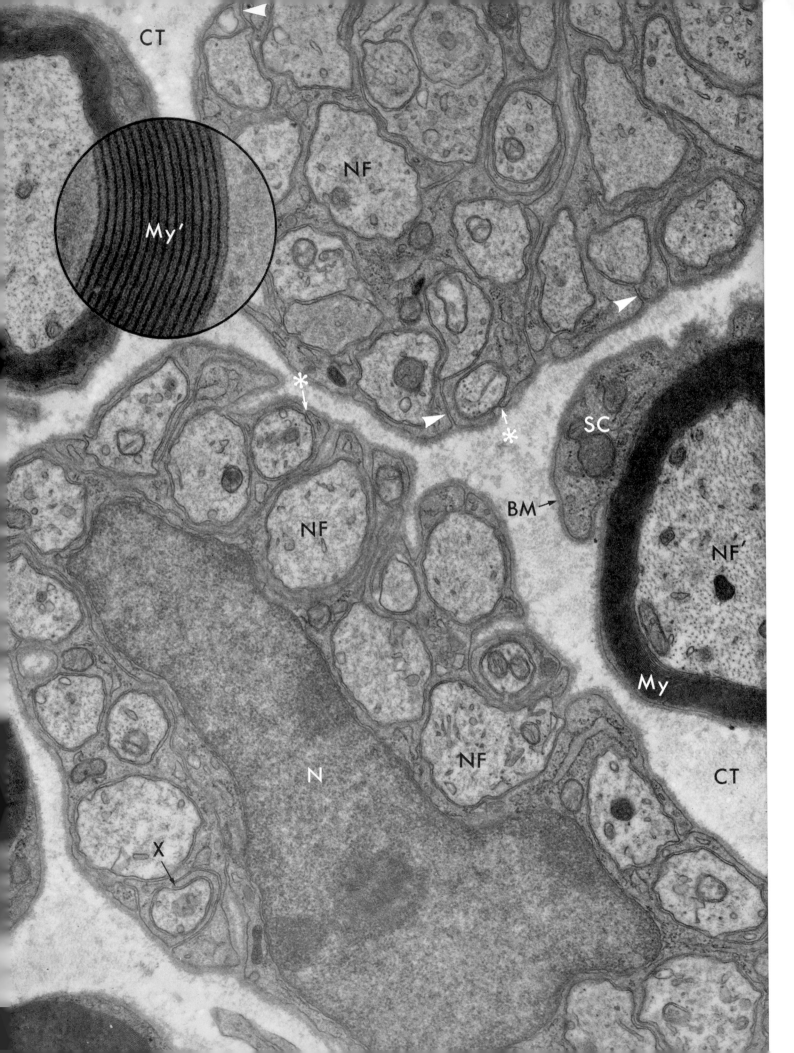

PLATE 44

Peripheral Nerve Fibers

THE long processes (fibers) of neurons are bundled together to form the "nerves" of gross anatomy. In cross sections of such nerves, as shown in this micrograph, the relationships of the fibers to special sheaths, coverings, and supporting connective tissue can be examined. Thus it has been learned that small, slow-conducting nerve fibers are not "naked" as was formerly believed but rather are enclosed by the cytoplasm of Schwann cells. Cross sections of many nerve fibers (NF) may be disposed around the nucleus (N) of a single Schwann cell, each enveloped by cytoplasmic extensions of the sheath cell. In favorable areas, the plasma membrane of the Schwann cell can be traced without discontinuity as it folds in from the surface and surrounds the fiber (∗). Where the fiber is deeply embedded in the Schwann cell, a narrow channel is defined by the apposition of lips of Schwann cell cytoplasm (arrows). The channel and membranes facing it constitute the mesaxon.

Larger nerve fibers with faster conduction rates (NF′) are covered by a myelin sheath (My). The formation of the sheath and its relationship to the nerve fiber has in recent years been greatly clarified by fine structural studies. Briefly, it has been found that the lamellae of the sheath (see inset, My′) are derived from successive layers of Schwann cell plasma membrane. In this micrograph, the image at X could represent an early phase in the development of this sheath. As myelination proceeds, a lip of cytoplasm, such as that at the point of the arrow, extends itself around the axon by synthesizing membrane as it goes. Behind this lip, the cytoplasm is squeezed out to the point where the two membranes of the Schwann cell behind the lip come together and fuse at their inner (cytoplasmic) surfaces (see text figure 44a). This fusion forms the strong or major dense line of the sheath image (inset). The space between the lines represents the face-to-face contact of two plasma membranes, in which contact the outer leaflet of the unit membrane loses its clear identity (see text figure 44a). By this progressive membrane growth at the margins of the cytoplasmic lips, the mesaxon is greatly extended, and the resultant sheath becomes a multilayered structure. Thus it is evident that the large nerve fibers, like the smaller ones, are enclosed by a living cellular sheath. Portions of the enveloping Schwann cell (SC) may be identified at the periphery of the sheath. Each Schwann cell is surrounded by an amorphous basement membrane (BM) that separates it from the connective tissue of the endoneurium (CT).

The evenly spaced, layered structure of myelin described above can be related to the known chemical composition and the arrangement of the macromolecules represented. As already pointed out, it has been learned from studies on the development of myelin that each major period, i.e., from center to center of each dense line (∼120 A) represents two unit membranes in face-to-face compression. The sheath is therefore pure membrane, which upon analysis is found to contain phospholipids, proteins, polysaccharides, salts and water. X-ray diffraction studies, combined with electron microscopy, have revealed that the low density regions represent the long hydrocarbon tails of the lipid molecules, regularly arranged and oriented perpendicular to the plane of the membrane. The denser bands, on the other hand, represent the polar groups of the lipid in combination with proteins. Why these combine on the back face of the unit membrane to give the major dense band in the myelin as opposed to the thinner and lighter line at the other face has not been explained.

From the skin of the mouse
Magnification × 27,500
Inset × 115,000

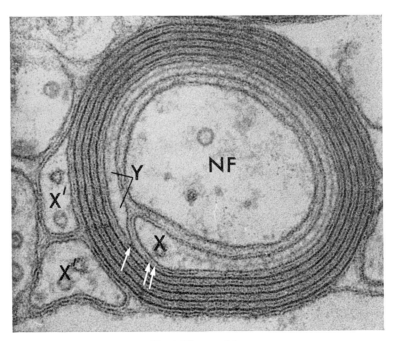

Text Figure 44a

This excellent micrograph shows clearly the structure of the myelin sheath and illustrates how the layers are formed within it as the neuroglial cell, the counterpart of the Schwann cell in the central nervous system, envelops the axon. The growing lip of cell cytoplasm (X) is advancing around the axon process (NF) and insinuating itself into the space between the plasma membrane of the axon and the membrane that limits the thin layer of cytoplasm (Y) left behind by the growing lip during its previous turn. This cytoplasmic layer disappears as the inner leaflets of its plasma membrane fuse to form the major dense line of the myelin sheath. This process is occurring at the point indicated by the single arrow. The outer leaflet of the plasma membrane surrounding the lip fuses with its own outer leaflet laid down on the previous turn. The two outer leaflets thus give rise to the less dense intermediate line of the sheath (double arrow). The cell body from which the investing cytoplasmic sheet originated cannot be seen in this micrograph, but cytoplasm within the lateral margins of the sheet do appear (X'). This micrograph was generously provided by A. Hirano and H. M. Dembritzer. It originally appeared in *J. Cell Biol., 34:555* (1967), where a more complete explanation of myelin sheath formation is provided.

From the brain of the rat
Magnification × 166,000

References

(*See also References to Plate 45*)

BUNGE, M. B., BUNGE, R. P., and RIS, H. Ultrastructural study of remyelination in an experimental lesion in adult cat spinal cord. *J. Biophysic. and Biochem. Cytol., 10:*67 (1961).

ELFVIN, L.-G. Electron microscopic investigation of the plasma membrane and myelin sheath of autonomic nerve fibers in the cat. *J. Ultrastruct. Res., 5:*388 (1961).

FERNÁNDEZ-MORÁN, H., and FINEAN, J. B. Electron microscope and low-angle X-ray diffraction studies of the nerve myelin sheath. *J. Biophysic. and Biochem. Cytol., 3:*725 (1957).

GASSER, H. S. Properties of dorsal root unmedullated fibers on the two sides of the ganglion. *J. Gen. Physiol., 38:*709 (1955).

GEREN, B. B. The formation from the Schwann cell surface of myelin in the peripheral nerves of chick embryos. *Exp. Cell Res., 7:*558 (1954).

HIRANO, A., and DEMBITZER, H. M. A structural analysis of the myelin sheath in the central nervous system. *J. Cell Biol., 34:*555 (1967).

MATURANA, H. R. The fine anatomy of the optic nerve of anurans—an electron microscope study. *J. Biophysic. and Biochem. Cytol., 7:*107 (1960).

O'BRIEN, J. S. Stability of the myelin membrane. *Science, 147:*1099 (1965).

PETERS, A. The formation and structure of myelin sheaths in the central nervous system. *J. Biophysic. and Biochem. Cytol., 8:*431 (1960).

———, and VAUGHN, J. E. Microtubules and filaments in the axons and astrocytes of early postnatal rat optic nerves. *J. Cell Biol., 32:*113 (1967).

PLATE 45

The Node of Ranvier

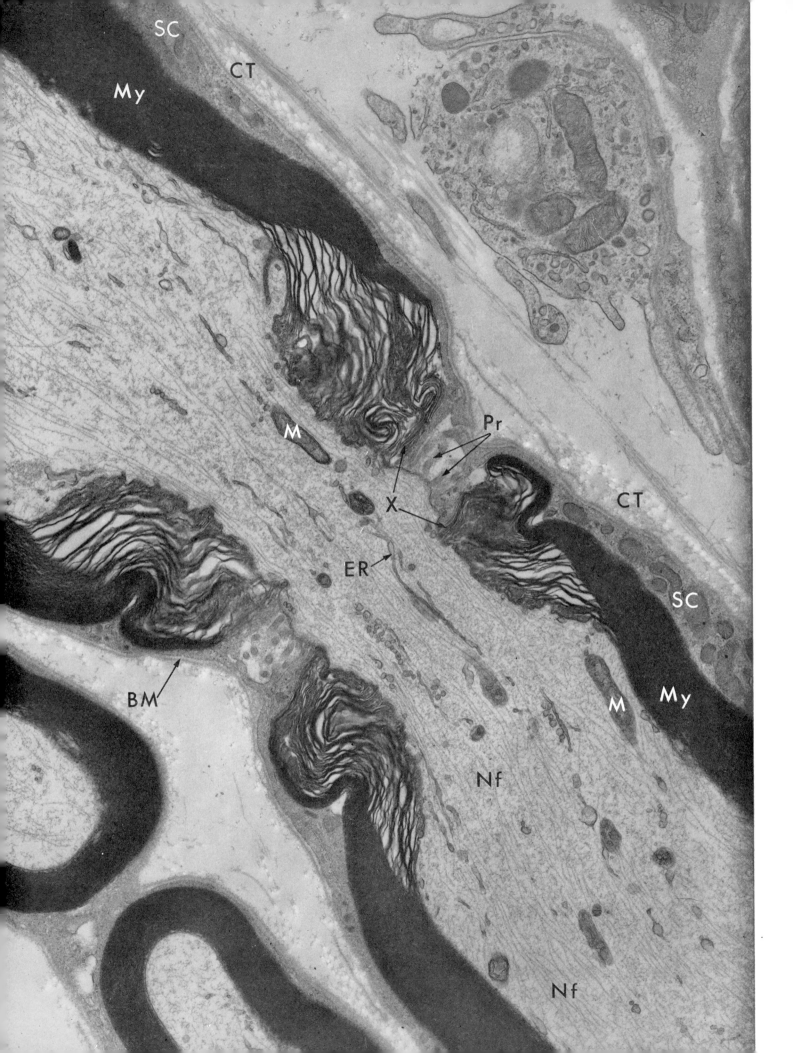

PLATE 45

The Node of Ranvier

THE myelin sheath of peripheral nerve processes (fibers) is periodically interrupted, and the gaps in this sheath are referred to as the nodes of Ranvier. Subsequent to understanding the nature of the myelin sheath (see Plate 44) we have come to learn more about this nodal structure. It has been found, for example, that a single Schwann cell is associated with each segment of the nerve fiber. The sheath of myelin, resulting from successive layering of Schwann cell plasma membranes, forms a compact tube (My) over most of the internodal areas. Near the node, some Schwann cell cytoplasm remains in the extended margins of the sheath layers and occupies a series of liplike folds (X), which envelop the fiber. In the region of the node itself, however, only fingerlike processes (Pr) of neighboring Schwann cells (SC) interdigitate and cover the nodal area. A basement membrane (BM) and connective tissue fibers (CT) of the endoneurium complete the wrappings of the fiber.

What is the significance of the nodes? An answer has been provided, based on the observation that the speed of conduction of myelinated fibers exceeds that of non-myelinated ones. According to a theory now widely favored, depolarization of myelinated nerve fibers occurs only in the region of the node, where the lipoprotein sheath of myelin, acting as an insulator, is absent. Thus the current can flow only in the nodal areas, and the impulse "jumps" from node to node. This phenomenon is called saltatory conduction. Good correlation has therefore been obtained between the fine structure of the node, observed by electron microscopy, and a large accumulation of physiological data.

While the primary events of impulse conduction are associated with the axonal plasma membrane, the membrane can function over a long period only when it is the limiting layer for a living core of axoplasm. The latter, it may be noted, is rich in neurofilaments (Nf) and contains slender elements of the ER and small numbers of thin mitochondria (M). Of special interest are the neurotubules, really microtubules (see text figure 45a), because these structures are probably involved in flow of protoplasm along the axonal fiber. The movement of protoplasm distally from the cell body has been documented by a number of work-

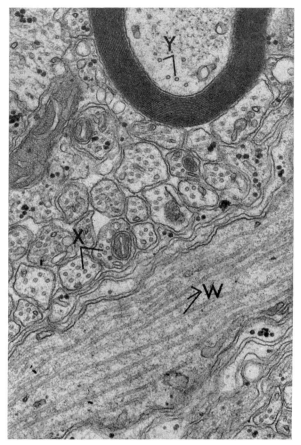

Text Figure 45a

This thin section of frog cerebellum demonstrates the presence of microtubules or, as they are called in this tissue, neurotubules, in the axoplasm of neurons. Profiles of tubules cut in cross section are found both in the small, nonmyelinated (X) nerve processes (dendrites) and in larger myelinated ones (Y) (axons). Longitudinally sectioned neurotubules (W) are present in the dendrite running diagonally across the lower part of the field.

From the cerebellum of the frog
Magnification × 36,000

ers. Although estimates of flow rate have varied, in general they resolve themselves into two: one about 1.2 mm per day and another of the order of 100 to 200 mm per day. Presumably the former represents a growth rate and the latter the protoplasmic streaming that might carry formed structures down the axon. Neither of these could easily provide for the local metabolic needs of the axon at all levels along its length, and so it

185

remains reasonable to look for other sources. Recent autoradiographic studies, tracing the incorporation of labeled amino acid into neurons, indicate that molecules may pass through the myelin sheath and enter the nerve cell process at points far removed from the nerve cell body. While the role of the sheath as a kind of insulator in the internodal regions has been noted (see above), attention is directed to the possibility that the sheath provides at the local level specific nutrients to the axon.

From the sciatic nerve of the mouse
Magnification $\times$ 22,500

References

BUNGE, R. P. Glial cells and the central myelin sheath. *Physiol. Rev., 48:*197 (1968).

CASTON, J. D., and SINGER, M. Amino acid uptake and incorporation into macromolecules of peripheral nerves. *J. Neurochem., 16:*1309 (1969).

ELFVIN, L.-G. The ultrastructure of the nodes of Ranvier in cat sympathetic nerve fibers. *J. Ultrastruct. Res., 5:* 374 (1961).

FRIEDE, R. L., and SAMORAJSKI, T. Myelin formation in the sciatic nerve of the rat. A quantitative electron microscopic, histochemical and radioautographic study. *J. Neuropathol. Exp. Neurol., 27:*546 (1968).

————, and SAMORAJSKI, T. The clefts of Schmidt-Lantermann: a quantitative electron microscopic study of their structure in developing and adult sciatic nerves of the rat. *Anat. Rec., 165:*89 (1969).

HEDLEY-WHYTE, E. T., RAWLINS, F. A., SALPETER, M. M., and UZMAN, B. G. Distribution of cholesterol-I, 2-H³ during maturation of mouse peripheral nerve. *Lab. Invest., 21:*536 (1969).

HENDELMAN, W. J., and BUNGE, R. P. Radioautographic studies of choline incorporation into peripheral nerve myelin. *J. Cell Biol., 40:*190 (1969).

NAPOLITANO, L. M., and SCALLEN, T. J. Observations on the fine structure of peripheral nerve myelin. *Anat. Rec., 163:*1 (1969).

REVEL, J.-P., and HAMILTON, D. W. The double nature of the intermediate dense line in peripheral nerve myelin. *Anat. Rec., 163:*7 (1969).

ROBERTSON, J. D. Preliminary observations on the ultrastructure of nodes of Ranvier. *Z. Zellforsch., 50:*553 (1959).

SINGER, M., and SALPETER, M. M. The transport of ³H-l-histidine through the Schwann and myelin sheath into the axon, including a re-evaluation of myelin function. *J. Morph., 120:*281 (1966).

UZMAN, B. G., and NOGUEIRA-GRAF, G. Electron microscope studies of the formation of nodes of Ranvier in mouse sciatic nerves. *J. Biophysic. and Biochem. Cytol., 3:*589 (1957).

WEBSTER, H. DEF. The geometry of peripheral myelin sheaths during their formation and growth in rat sciatic nerves. *J. Cell Biol., 48:*348 (1971).

WEISS, P., and HISCOE, H. B. Experiments on the mechanism of nerve growth. *J. Exp. Zool., 107:*315 (1948).

PLATE 46

The Rod Outer Segment

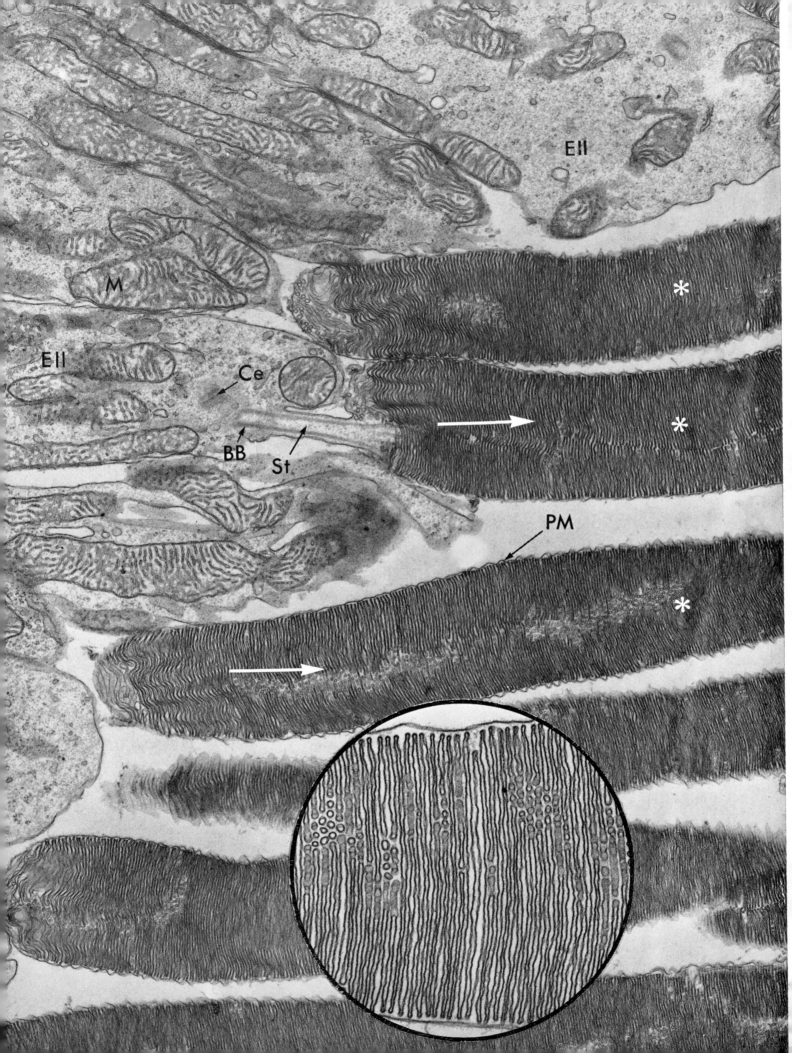

PLATE 46

The Rod Outer Segment

The retina is a photosensitive epithelium which, when excited by light, is able to transmit to the brain impulses that are interpreted as visual images. The epithelium consists of two types of bipolar neurons, which are named for the shapes of their sensory tips or outer segments, that is, the rods and cones. Rods are sensitive to dimmer illumination, but unlike the cones, which require more light for excitation, they cannot perceive color. The neuron cell bodies, from which the outer segments project, harbor the cell nucleus together with cytoplasmic organelles such as ribosomes and mitochondria. An axon extends from the pole of each cell opposite the sensory ending.

In this micrograph only parts of six rod outer segments (∗) and portions of neuronal cytoplasm, the so-called ellipsoids (Ell), are seen. These accommodate a large population of mitochondria (M). Other parts of the rod cell extend out of the picture to the left. Cones, on the other hand, are absent from the retina of the nocturnal kangaroo rat, the source of this specimen. However, we can observe a number of interesting structural features of the rods and attempt to evaluate their functional significance.

Rod outer segments are slender, cylindrical structures about 1.2 micrometers in diameter and 15 to 20 micrometers long. Their limiting membrane (PM) is continuous with that of the rest of the rod cell by way of a slender connecting stalk (St). The latter has the structure of a cilium (see Plate 8) and extends from a basal body (BB). Nearby one can identify part of the centriole (Ce) that is constantly associated with the basal body. The remarkable feature of these rod outer segments, which was early revealed by electron microscopy, is the laminated structure of their contents. They are essentially stacks of thin, membranous sacs (see inset). These are shaped like coins or disks with, however, some central perforations. In some species the margins of the lamellae are indented to form fissures, and thus in cross sections of the rod the disks have a scalloped profile.

Since the visual pigments are known to reside in the outer segments, the layered membranes provide strata in or on which pigments are oriented with respect to impinging light rays. These, by virture of the structure of the eye, follow a path along the long axes of the outer segments (in the direction of the white arrows). Meticulous study of the light absorbing properties of small pieces of the retina and also of individual cells therein has indicated that the rod-shaped pigment molecules are located transversely with respect to the long axis of the segment and are therefore in a position to absorb maximally.

When the membranes of the disks are examined in face view at high magnifications with the electron microscope, small circular particles (30 nm in diameter) may be found evenly distributed over them. These particles have also been described as disks with about the same thickness as the membrane (50 A). Although the micrographs cannot reveal their composition, it is reasonable to postulate that the visual pigments are concentrated in them. In support of this, calculations have shown that the amount of visual pigment in a rod could be included within these particles.

The visual pigments are made up of an aldehyde of vitamin A (a retinal) joined to a specific protein. The light in striking the pigment isomerizes the aldehyde, and this leads in turn to the separation of the retinal from the protein and to the exposure of presumably reactive groups on the protein. In some as yet unknown way these initial reactions lead to electrical excitation of the neuron and thereby give rise to impulses transmitted to the brain. It is thought that the excitation is elicited in the plasma membrane enclosing the disks of the rod and from that area moves as a wave of depolarization over the cell body and axon. These phenomena are analagous to those encountered in excitation of neurons in general (see Plate 43).

The origin of membranes in the outer segment of the rods (and cones) and indeed of the segments themselves has intrigued cytologists. Here as in other sensory epithelia (see Plate 47) the receptor ending seems to be a modified cilium. The disks, however, do not arise from ciliary microtubules but rather as infoldings of the plasma membrane at the bases of the outer segments. This is true of both cones and rods, although it is only in the latter that the membranes lose their continuity with the plasma membrane so that they encompass a closed space. This spectacular production of membrane depends in some way upon the presence of the cilium, or at least the basal body, and also upon the presence of vitamin A. Vitamin A deficiency causes progressive degeneration of the outer segments, but recovery is possible when vitamin A is supplied and when the basal body has not been destroyed. Recently it has been found that normally the disks are being continuously renewed even in the mature animal, with new ones presumably being added at the base of the outer segment.

The retina of the kangaroo rat (*Dipodomys ordi*)
Magnification × 30,000
Inset × 59,000

References

BROWN, P. K., GIBBONS, I. R., and WALD, G. The visual cells and visual pigment of the mudpuppy, *Necturus*. *J. Cell Biol., 19:*79 (1963).

COHEN, A. I. Electron microscope observations on form changes in photoreceptor outer segments and their saccules in response to osmotic stress. *J. Cell Biol., 48:*547 (1971).

DEROBERTIS, E. Electron microscope observations on the submicroscopic organization of the retinal rods. *J. Biophysic. and Biochem. Cytol., 2:*319 (1956).

DOWLING, J. E. Night blindness. *Sci. Amer., 215:*78 (October, 1966).

HUBBARD, R., and KROPF, A. Molecular isomers in vision. *Sci. Amer., 216:*64 (June, 1967).

KORENBROT, J. I., BROWN, D. T., and CONE, R. A. Membrane characteristics and osmotic behavior of isolated rod outer segments. *J. Cell Biol., 56:*389 (1973).

LAM, D. M. K. The biosynthesis and content of gamma-aminobutyric acid in the goldfish retina. *J. Cell Biol., 54:*225 (1972).

LATIES, A. M., and LIEBMAN, P. A. Cones of living amphibian eye: selective staining. *Science, 168:*1475 (1970).

NILSSON, S. E. G. Receptor cell outer segment development and ultrastructure of disk membranes in the retina of the tadpole (*Rana pipiens*). *J. Ultrastruct. Res., 11:*581 (1964).

SJÖSTRAND, F. S. Electron microscopy of the retina. *In* The Structure of the Eye. G. K. Smelser, editor. New York, Academic Press (1961) p. 1.

WALD, G. General discussion of retinal structure in relation to the visual process. *In* The Structure of the Eye. G. K. Smelser, editor. New York, Academic Press (1961) p. 101.

WEBB, N. G. X-ray diffraction from outer segments of visual cells in intact eyes of the frog. *Nature, 235:*44 (1972).

YOUNG, R. W. The renewal of photoreceptor cell outer segments. *J. Cell Biol., 33:*61 (1967).

190

PLATE 47

The Olfactory Epithelium

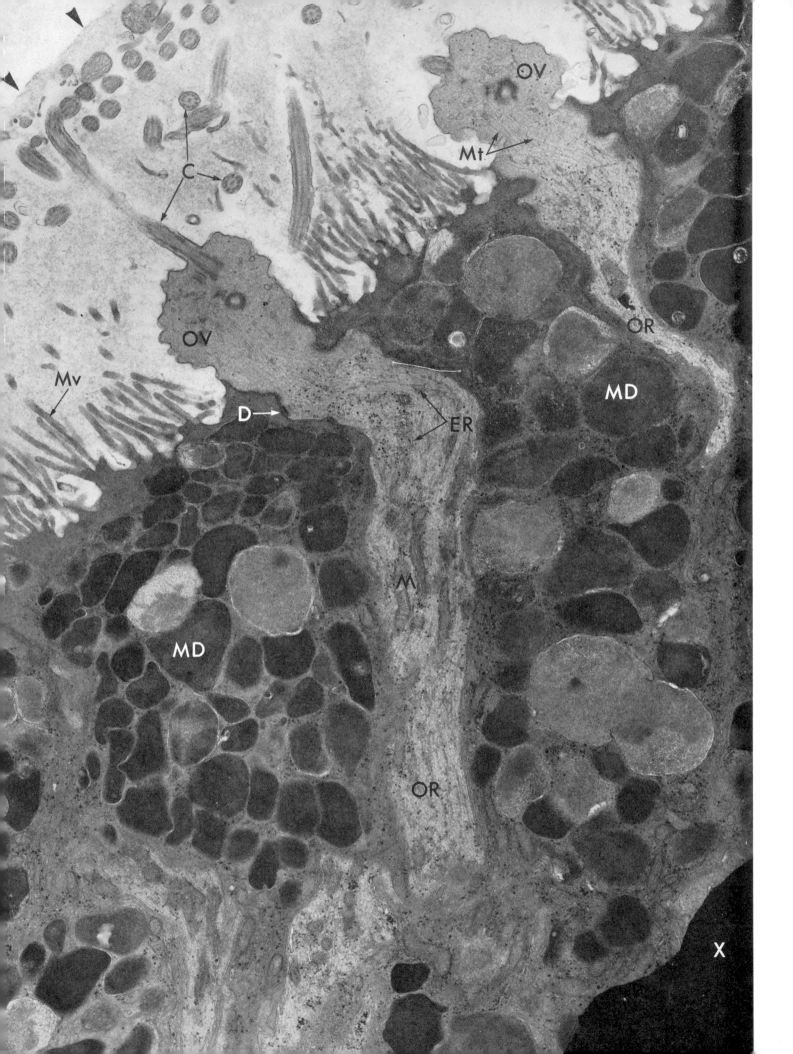

PLATE 47

The Olfactory Epithelium

THE ability to discriminate among numerous volatile components of the environment depends on the functioning of a pseudostratified epithelium that covers the roof of the nasal cavities. The receptor cells of this epithelium are extraordinarily sensitive to the quality and quantity of the stimulant that can be detected. The mechanism by which this remarkable discrimination is achieved is not known; indeed, it is not absolutely certain what structure or structures of the receptor cell surfaces are involved. All that is known is that certain bipolar neurons in the epithelium are differentiated to respond to chemostimulation and that the discrimination is based on a primary reaction between the exposed cell surface (or its extensions) and the stimulating agent. It is the role, then, of the neuron to translate its reaction with the stimulant into a series of nerve impulses, which the brain can interpret as representing a certain substance in strong or weak concentrations.

When these facts are considered, it is in a way surprising to find that the cell doing this complex job appears relatively simple. Its central cytoplasmic mass containing the nucleus resides at a level midway between the apical and basal surfaces of the epithelium (text figure 47a). From the distal pole of this cell a slender process, called the olfactory rod, runs to the free surface of the epithelium within crevices between the surrounding sustentacular cells. It may be attached to the latter by desmosomes (D). At its extreme tip this dendritic process projects slightly beyond the epithelial surface to form a bulbous ending, which represents the part of the cell exposed for chemoreception. From the other end of the cell body an even more slender process, the axonal fiber, extends out of the epithelium and into the lamina propria where it joins others to make up the olfactory nerve. Each receptor cell with its axon functions as an independent channel carrying information to the higher coordinating centers of the brain.

The other cells, comprising the bulk of the olfactory epithelium, have long been referred to as sustentacular because they surround the bipolar neurons and maintain the integrity of the epithelium. Since these cells produce in part the mucous layer that covers the epithelium, they have a synthetic and secretory function as well. The nuclei of these tall columnar cells are located generally at a level above those of the neurons (text figure 47a).

It is the purpose of the accompanying plate to show the most interesting and unusual part of this epithelium, the free surface where chemoreception takes place. Thus the field of the micrograph includes the apical poles of a few sustentacular cells with large mucous droplets (MD) and the olfactory rods of three receptor cells (OR). Two of these rods extend beyond the free surface of the epithelium and end in bulbous protrusions, inappropriately called the olfactory vesicles (OV). The cytoplasm of these nerve cells is populated by structures commonly found in dendritic processes. Mitochondria (M) are slender and tend to be crowded into the zone just below the level of the free surface. The dense particulate material is probably glycogen. There are a few slender profiles of the endoplasmic reticulum. And finally there are numerous neurotubules (Mt), about 200 A in diameter, which may play a role in the to-and-fro flow of the cytoplasm, i.e., the ground substance, in these slender cell extensions.

The free surfaces of the sustentacular and the receptor cells are seen to differ strikingly. From the former, long slender microvilli (Mv), 50 nm in diameter, project into the thick mucous layer. The bulbous dendritic endings (OV) lack these but have instead several cilia (C), 10 to 20 per cell, identified readily as such by the 9 + 2 complex evident in cross sections (see Plate 8). The cilia start from basal bodies in the cell cortex and radiate out in all directions from the crown of the bulb. Some of them are extremely long (100 to 200 μm), whereas others are reported to be much shorter (20 μm) and motile. The longer ones extend to the limits of the mucous layer (arrows), bend sharply into the plane of this surface, and become part of a network of fine ciliary endings. The tips are less than half the diameter of their bases, and the 9 + 2 complex evident in the basal part of the cilium is replaced by an unorganized bundle of microtubules.

There is general agreement among physiologists that these slender cilia and their distal ends, located just within the surface of the mucous layer,

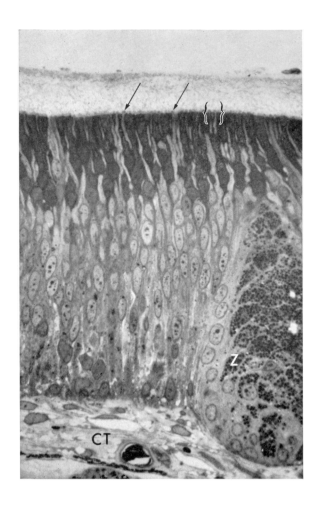

Text Figure 47a

This photomicrograph depicts the microscopic anatomy of the entire thickness of the olfactory epithelium. It was made from a ½ μm section stained with toluidine blue. The slender lightly staining elements that reach up to and beyond the free surface of the epithelium (arrows) represent the olfactory rods of the bipolar receptor cells. They are sandwiched between the more deeply stained sustentacular cells, which make up the bulk of the upper third of this pseudostratified epithelium. Slender processes extend from the sustentaculars to the basement membrane. It is obvious that the nuclei and cell bodies of the bipolars reside below those of the sustentaculars. A layer of connective tissue (CT) with vascular elements underlies the epithelium. The tissue at the right (Z) is part of a Bowman's gland, which produces and secretes mucous material for the outer layer in which the cilia (faintly shown) are embedded. This layer in this preparation is about 20 μm thick.

The electron micrograph in Plate 47 includes only a small area of the epithelium such as that enclosed by brackets.

This photomicrograph was kindly provided by Dr. T. S. Reese.

Magnification × ca. 600

---

are the structures primarily sensitive to odorous material. This induces us to ask how the nerve impulse is initiated and how discrimination is achieved. The theories proposed to account for these phenomena are too numerous to review here, and, in any case, the majority have not survived experimental tests. One idea currently popular with students of olfaction proposes that there are 7 to 10 primary odors and that these are referable to the geometric forms of the stimulating molecules rather than to details of their chemical composition or structure. The subtle psycho-

logical interpretations of, let us say, perfumes derive from different combinations of these 7 to 10 primary odors. Speculation on the involvement of cell structure in discrimination proposes that different receptor cells respond to different combinations of the primaries because of different ratios of primary receptor sites, represented possibly by different cilia and their slender terminals. (The dense area at X represents a bar of the supporting grid.)

From the olfactory epithelium of the leopard frog
Magnification × 20,000

AMOORE, J. E., JOHNSTON, J. W., JR., and RUBIN, M. The stereochemical theory of odor. *Sci. Amer., 210:*42 (February, 1964).

DE LORENZO, A. J. D. Studies on the ultrastructure and histophysiology of cell membranes, nerve fibers and synaptic junctions in chemoreceptors. *In* Olfaction and Taste. Wenner-Gren Center International Symposium Series. Y. Zotterman, editor. New York, The Macmillan Company, vol. I (1963) p. 5.

GASSER, H. S. Olfactory nerve fibers. *J. Gen. Physiol., 39:*473 (1955).

GESTELAND, R. C., LETTVIN, J. Y., PITTS, W. H., and ROJAS, A. Odor specificities of the frog's olfactory receptors. *In* Olfaction and Taste. Wenner-Gren Center International Symposium Series. Y. Zotterman, editor. New York, The Macmillan Company, vol. I (1963) p. 19.

LE GROS CLARK, W. Inquiries into the anatomical basis of olfactory discrimination. *Proc. Roy. Soc. London, Ser. B., 146:*299 (1957).

OKANO, M., WEBER, A. F., and FROMMES, S. P. Electron microscopic studies of the distal border of the canine olfactory epithelium. *J. Ultrastruct. Res., 17:*487 (1967).

OTTOSON, D. Generation and transmission of signals in the olfactory system. *In* Olfaction and Taste. Wenner-Gren Center International Symposium Series. Y. Zotterman, editor. New York, The Macmillan Company, vol. I (1963) p. 35.

REESE, T. S. Olfactory cilia in the frog. *J. Cell Biol., 25* (No. 2, part 2): 209 (1965).

Notes on Technique for Transmission
Electron Microscopy

CERTAIN features in the construction and operation of the electron microscope oblige the microscopist to use special techniques in the preparation of his specimens. Foremost among these is the high vacuum required in the column of the microscope during operation. Obviously, the cell or other specimen exposed to this vacuum cannot be kept alive or even hydrated under ordinary conditions of observation. It must instead be preserved or "fixed" in a form that as closely as possible resembles the morphology of the living state. Though a number of chemical reagents can be used for this fixation, none has greater over-all value than osmium tetroxide (OsO_4). In addition to preserving quite faithfully the form of the living material, it reacts differentially with cell components. The resulting uneven deposition of osmium atoms exaggerates density differences in the specimen and gives it the appearance of being stained. This effect is important in giving the image enough contrast to permit the observer to distinguish one structural component from another in areas where natural density differences would be insufficient. In spite of these demonstrable values of OsO_4, the search for better reagents continues. Investigators are currently finding formaldehyde and glutaraldehyde, followed by osmium tetroxide, more reliable for the preservation of the complete complement of materials making up the cell's fine structure.

Another problem of specimen preparation develops from the fact that the electron beam will penetrate only a very thin layer of organic material. To meet this limitation, the microscopist must prepare extraordinarily thin sections, not thicker than 0.05 to 0.1 micrometers. Therefore, in its preparation for microtomy the specimen is first dehydrated with alcohol and subsequently infiltrated with a resin or plastic in monomeric form. This, in turn, in polymerized by heat and catalysts, and the tissue, now embedded, is hard enough to section with a glass or diamond knife on microtomes designed especially for the task. The resulting thin sections may be given an additional staining with a salt of some heavy metal, such as lead or uranium, and the resulting specimen is ready for microscopy.

The micrographs originating in the Laboratory for Cell Biology at Harvard University were taken with a Philips 200 electron microscope (EM). This instrument is similar to others currently available in yielding resolutions at least 100 times greater than those given by the light microscope. Such resolutions make it feasible to take pictures with the EM at magnifications of 40 or 50 thousand times and to separate or resolve points of density as close together as 10 Angstrom units (10 A). For many purposes, however, especially for forming first impressions of the fine structure in a cell or tissue, lower magnifications are preferable. Thus, many of the originals of these pictures were taken at magnifications of only 3 to 5 thousand and thereafter were enlarged photographically.

With a knowledge of the magnification, one can, of course, measure the dimensions of objects in the pictures. This is commonly done in terms of small units of length: micrometers (μm), formerly called microns (μ); nanometers (nm), formerly called millimicrons ($m\mu$); and Angstrom units (A). In this connection the student will recall that

1 μm $=$ 1/1000 of a millimeter (mm)
1 nm $=$ 1/1000 of a micrometer (μm)
1 A $=$ 1/10 of a nanometer (nm)

It follows that in a micrograph having a magnification of 1000×, a μm length is represented by 1 mm on the picture, or that at 30,000×, 1 μm is represented by 30 mm. Fractions of these distances, as measured with a millimeter scale or a pair of calipers, can be readily translated into nm's or A's, and the dimensions of the object thereby determined.

197

Index

Numerals in *italics* refer to illustrations.